THE SCIENCE GAP

THE SCIENCE GAP

Dispelling the Myths and
Understanding the Reality of Science

MILTON A.
ROTHMAN

PROMETHEUS BOOKS
Buffalo • New York

Published 1992 by Prometheus Books

96 95 94 93 92   5 4 3 2 1

Library of Congress Cataloging-in-Publication Data

Rothman, Milton A.
   The science gap : dispelling the myths and understanding the reality of science /
by Milton A. Rothman.
       p.   cm.
   Includes bibliographical references and index.
   ISBN 0-87975-710-8
   1. Science—Philosophy. 2. Mythology. I. Title.
Q175.R5649   1992
500—dc20   91-27481
                                                                   CIP

Printed in the United States of America on acid-free paper.

Thanks to Miriam for patience and encouragement.

CONTENTS

INTRODUCTION

Myth n. Any fictitious story, or unscientific account, theory, belief, etc.

Two kinds of myths about science abound in the world. Myths of the first kind are theories which for a time are believed by large numbers of people, at least until they are eventually shown to be false. The second kind of myth is about the nature of science itself.

The most familiar example of a scientific theory that turned out to be a myth is the idea, once widely held, that the earth is the center of the solar system. Other important examples of past scientific myths are the notions that heat consists of a fluid called caloric; that combustion is due to another fluid called phlogiston; and that light consists of waves in a universal fluid called the ether. These ideas were part of established science until well into the nineteenth century, when the growth of modern physics pointed the way to more realistic concepts.

The idea that the action of the nervous system requires the presence of entities such as psychic energy and *elan vital* is still believed by many non-scientists, and even by some psychologists, but it is not accepted by any contemporary neuroscientists that I know of.

The fable that certain markings on Mars represent huge canals was originally promulgated by a few professional astronomers, and for a period was part of the mythology of science fiction. However, as soon as the Viking and Mariner spacecrafts were in a position to photograph the surface of Mars from close proximity (during the 1970s), there were suddenly no canals to be seen and the myth suffered total eclipse.

The myth that the positions of the planets determine the destiny of individuals on earth was once believed by scientists as prominent as Kepler and Newton, but is no longer a part of scientific knowledge, even though a large portion of the general public still subscribes to it under the rubric of "astrology." So pervasive is this myth that some journalists cannot tell the difference between astrology and astronomy.

The idea that the earth was created approximately ten thousand years ago, and that each species of animal was created for a specific purpose by a supernatural being is a myth vigorously put forward by a group of people who call themselves "creationists." There is not a professional biologist, paleontologist, or geologist who accepts this thesis as valid science. Yet many school board members with beliefs of a fundamentalist religious nature are working vigorously to get this theory taught in public schools on the same footing as the scientifically valid principle of evolution.

All of the above myths are theories which are not true in any sense that scientists understand, i.e., there is no physical evidence by which these theories may be validated, and there is an enormous amount of evidence that contradicts them. Nevertheless many people still believe some of these ideas to be true because they want to believe them, and because they have a need to believe them. That is the nature of myth.

Myths of the second kind are myths about the nature of science itself. Those are embodied in a number of common adages—folk sayings such as: "Nothing is known for sure," "Anything is possible," and "Any theories we believe today are likely to be overturned in the future." These statements are bandied about in conversation and in the popular press without a great deal of reflection as to their real meaning. They are false generalizations that make it easy for people to believe in myths of the first kind.

In this book I intend to deal primarily with myths of the latter kind, for there is already an ample literature that deals with myths of the first kind. The history of science is the history of theories which were believed for a time by the scientific community, but which were replaced eventually by better theories. Myths believed by the general public but never accepted by most scientists fall within the history of pseudoscience or paranormal phenomena.[1-9]

I include in this literature the classic work of the nineteenth century English mathematician, Augustus De Morgan. Even though his writing deals with pseudomathematics instead of pseudoscience, De Morgan deserves recognition as one of the first people to discuss seriously the characteristics of the ubiquitous science crank. This creature, who dwells on the fringes of the professional scientific community, was dubbed by De Morgan to be a *paradoxer,* an appellation more charitable than the more colloquial or pejorative term "crackpot." De Morgan used the term "paradox" in accordance with the first definition in the *Oxford English Dictionary:*

> A statement or tenet contrary to received opinion or belief; often with the implication that it is marvelous or incredible; sometimes with unfavourable connotation, as being discordant with what is held to be established truth and hence absurd or fantastic; sometimes with favourable connotation, as a correction of vulgar error. (In actual use rare since 17th c., though often insisted upon by writers as the proper sense.)

Thus, *paradox* is the opposite of *orthodox,* and in fact is seen to be identical with *paranormal.* A paradoxer, in this sense, is a person who devotes his life to the cultivation of pseudoscientific theories, which De Morgan also called "crotchets." Among favorite crotchets of De Morgan's subjects were the classical mathematical problems of "trisecting the angle," and "squaring the circle."

A Century of Science and Other Essays by John Fiske contains a chapter titled "Some Cranks and their Crotchets," an extension of De Morgan's lampoon of circle-squarers and angle-trisectors. Fiske also takes on sundry perpetual-motion enthusiasts, numerologists, hollow-earthers, and pyramidologists. L. Sprague de Camp likewise pursues classical cultists, charlatans, Kabbalists, and pseudoscientists in *The Ragged Edge of Science.* The other books listed at the end of this chapter analyze and dispute a horde of paranormal claims ranging from N-rays and spoon-bending to highly technical parapsychology experiments.

The reasons for expending so much energy in debunking the work of others are many. We seldom change the minds of the believers, but we hope to educate those who are not quite convinced—especially students. Also, we must confess that much of this disputation takes place for our own pleasure. Augustus De Morgan confesses that he may have a little of the spirit of Colonel Quagg, who expressed his principle of action thus: "I licks ye because I kin, and because I like, and because ye's critters that licks is good for!"

There is little organized literature concerning myths of the second kind—myths about the nature of science itself—although I have referred to them tangentially in two previous works.[10–11] Consequently the present volume attempts to fill a gap in the literature of pseudoscience, with some forays into philosophy of science.

I write from the point of view of an experimental physicist with ten years of nuclear physics research, ten years in fusion energy research, and ten more years in teaching. My attitude has been consistently skeptical. A lifelong association with the science fiction community and several years association with the Committee for the Scientific Investigation of Claims of the Paranormal (CSICOP) have given me familiarity with the type of logic characteristic of those people who are inclined to believe in currently popular crotchets: astrology, ESP, parapsychology, UFOs, faster-than-light travel or communication, creationism, etc.

My viewpoint is rigorously realistic, though modified by the requirements of quantum mechanics. It is pragmatic in the sense that physical observation is the final arbiter of theory. I plead guilty to being reductionist, in the physical sense to be explained later in Myth 15. This leads to questions: Is physics all there is? Or is there more to the universe than physics? Do there exist forces and entities which do not fit into the scheme of science? ("Are there more things on heaven and earth . . . ?") We cannot say no with total certainty.

Nevertheless, as we survey the vast universe with all the physical instruments at our disposal, probing the minute space within each atom and observing what happens in the enormous regions beyond our galaxy, we find no objects behaving in a manner not described by the laws of physics or moving in response to forces that cannot be detected by ordinary physical means. Physics experiments give no hint of forces in nature outside of those observable in the physics laboratory or in stellar objects. In short, we find no evidence for the supernatural.

It is when we try to explain the processes of life—our awareness of being and our sense of controlling our own actions—that the mysteries become most profound. The more closely we look at what happens within our nervous systems, the less evidence we see of abstractions such as psychic energy, *elan vital,* and "mind" as an entity separate from the body. The more we turn up the magnification, the less we see of the ghost in the machine. We see nothing but atoms and molecules, doing what they have to do. The more we see, the less we understand.

But we have just started to look. Surely the more we learn, the more we will understand. This has been the trend for the past century.

One thing is clear: science and the supernatural inhabit two incompatable domains. Science includes everything observable by physical means; the supernatural deals with things not observable by any means known to science. Religion has to do with beliefs and feelings about supernatural entities, and so occupies a world of discourse separate from the world of science.

Most of the people in the world believe in the existence of entities and forces not recognized by science. This includes a large number of scientists who subscribe to religious beliefs and follow the ceremonies and traditions of religious institutions. It is possible for a well-functioning scientist to believe in the efficacy of prayer and—if he is sufficiently orthodox—to decline the use of the telephone on the sabbath. This is a situation fraught with conflict, not only within the minds of individual scientists, but in society as a whole. The following symptoms illustrate the nature of this conflict.

Even as our knowledge of nature has increased many-fold over the past two centuries, opposition to the natural sciences continues to rise. Certain conservative religious organizations attempt to control the content of textbooks and to limit what the public schools may teach about evolution or about human reproduction. A successful and popular book, *The Closing of the American Mind,* criticizes modern education and glorifies the teachings of Plato.[12] The author, Allan Bloom, would like modern students to read the Greek philosophers in their original form and in a serious manner. However, he fails to explain how those students are to obtain a realistic understanding of the world from authors who knew nothing about the true nature of the universe, and whose ideas about the universe hindered the development of modern science for hundreds of years. Professor Bloom deplores the current anarchy in the humanities. We scientists simultaneously deplore the universal decline of elementary education in the natural sciences.

The increase in knowledge among scientists has had little or no effect on the spread of irrationality among non-scientists. A typical bookstore has a few shelves of science books in one section and many more shelves of books on astrology, occultism, and assorted paranormal phenomena in another section, even though no properly conducted scientific investigation has demonstrated a particle of merit in any claim of the supernatural or paranormal.[13] Even some books purporting to deliver hard science are in fact not-so-subtle attempts to introduce connections between quantum theory and mysticism[14] to the outrage of the working scientists who know the limitations of quantum physics. The result is an ever-increasing "science gap" between those whose knowledge of nature is based on reality and the large numbers who continue to rely on myth and mysticism for a view of the world.

My insistence that all theories about nature must meet the tests required of any scientific theory is sometimes greeted with the derisive charge that this attitude represents "scientism," and is, therefore, wrong.

> **Scientism** n. 1. the techniques, beliefs, or attitudes characteristic of scientists; 2. the principle that scientific methods can and should be applied in all fields of investigation; often a disparaging usage.[15]

I find it odd that a word used to describe what scientists do has become pejorative. It is true that scientific methods do not work well when we try to deal with matters outside the domain of science: aesthetics, ethics, literary criticism, and the like. But in dealing with the reality of nature, scientism is what scientists do. There is no scientific treatise, textbook, or journal that does otherwise. Any other path leads in the direction of fantasy and unreality.

The progress of science since Galileo—particularly in the past century—has been away from mystical ideas about things with unobservable properties. (The fundamental particle known as the quark, though unobservable as an isolated object, is a realistic construct because quark theory predicts observable and measureable effects.) This is the fruitful way to go, and critics using the term "scientism" as a weapon of debate have the obligation to spell out the real reasons behind their objections and to give us an alternative method of investigating nature.

This book is an expansion of an article that appeared in the *Skeptical Inquirer,* the journal of the *Committee for the Scientific Investigation of Claims of the Paranormal.*[16] It is intended for all those with an interest in science, the philosophy of science, and skepticism. I also hope that the book will be a valuable resource for educators and students, for they are the ones occupying the front lines in the war against irrationality and unreality. Technical details are kept to a necessary minimum, having been presented more fully in two previous books.[17] The two prior works together with the present volume constitute a trilogy, progressing from the detailed and physical, to the general and philosophical. If in this book I refer frequently to the two prior volumes, it is simply to avoid excessive repetition of material already given in some detail.

NOTES

1. A. De Morgan, *A Budget of Paradoxes,* 2nd ed. (Chicago: Open Court Publishing Co., 1915). (First edition, 1872.)

2. J. Fiske, *A Century of Science and Other Essays* (Boston and New York: Houghton, Mifflin and Company, 1899).

3. M. Gardner, *Fads and Fallacies in the Name of Science* (New York: Dover Publications, 1957).

4. M. Gardner, *Science: Good, Bad and Bogus* (Buffalo, N.Y.: Prometheus Books, 1981).

5. L. S. de Camp, *The Ragged Edge of Science* (Philadelphia: Owlswick Press, 1980).

6. T. Hines, *Pseudoscience and the Paranormal* (Buffalo, N.Y.: Prometheus Books, 1988).

7. G. O. Abell and B. Singer, eds., *Science and the Paranormal* (New York: Charles Scribner's Sons, 1983).

8. P. Kurtz, ed., *A Skeptic's Handbook of Parapsychology* (Buffalo, N.Y.: Prometheus Books, 1985).

9. K. Frazier, ed., *The Skeptical Inquirer,* Journal of The Committee for the Scientific Investigation of Claims of the Paranormal, Buffalo, N.Y.

10. M. A. Rothman, *Discovering the Natural Laws* (New York: Doubleday & Co., 1972, and New York: Dover Publications, 1989).

11. M. A. Rothman, *A Physicist's Guide to Skepticism* (Buffalo, N.Y.: Prometheus Books, 1988).

12. A. Bloom, *The Closing of the American Mind* (New York: Simon & Schuster, 1987).

13. See P. Kurtz, ed., *A Skeptic's Handbook.*

14. See J. Fiske, *A Century of Science and Other Essays.*

15. All definitions in this book are from *Webster's New World Dictionary,* Third College Edition, V. Neufeldt, ed. (New York: Webster's New World, 1988).

16. M. A. Rothman, "Myths about Science and Belief in the Paranormal," in *The Skeptical Inquirer* (Fall 1989): p. 25.

17. See M. A. Rothman, *Discovering the Natural Laws* and *A Physicist's Guide to Skepticism.*

MYTH 1

"Nothing exists until it is observed."

Setting the Stage: Metaphysical Idealism or Metaphysical Realism?

One of the most pervasive myths of science is the idea that "nothing is known for sure." This adage projects a disarming modesty. Even scientists have been known to fall into the arms of this myth in order to avoid the appearance of dogmatism and arrogance. But is it really true that nothing is known for sure? To experimental physicists and engineers the answer is: of course not! If they were not sure that elementary particles existed, they would not strain their intellects to build instruments designed to detect these particles and to measure their properties with the utmost accuracy. If they were not sure that electromagnetic waves existed, they would not build radio and TV receivers to catch sounds and sights from around the world. Then why do people continue to perpetuate the myth that "nothing is known for sure?"

This question brings us face to face with the foundations of classical philosophy. The experimental physicist has a point of view that might be called "naive realism," the idea that things do exist out there, things that can be observed and measured. However, there are other possible models of the universe. Historically, there have been two major ways of dealing with the question of existence and reality: metaphysical idealism and metaphysical realism. Of course, as is true in any philosophical dispute, there are many fine distinctions and points of view within each of the two major categories. But the basic differences between the idealist and the realist points of view are crucial.

Idealists (in their purest form) believe that things exist only as ideas in the mind as opposed to material objects independent of the mind. The world outside of the mind is generated by extrapolating from these thoughts, but there is no way of proving that the outside world really exists. Idealists of a less extreme persuasion admit that there is a universe outside the mind, but that the mind is necessary for its existence or for our perception of its nature.

Implicit in idealist philosophy is the concept of *mind* as a fundamental entity; the body is separate from the mind and is not necessary to the theory of reality. Mind takes precedence over matter. The tradition behind idealist theory is long. The Platonists of Greece rejected the outer, real world in favor of the self-created inner world of pure thought.[1] The Platonists, according to Bertrand Russell, believed that "opinion is of the world presented to senses, whereas knowledge is of a super-sensible eternal world; for instance, opinion is concerned with particular beautiful things, but knowledge is concerned with beauty in itself."[2] But opinion is changeable. Thus the senses provide us with uncertain knowledge. True knowledge of the eternal world can be obtained only by reason, intelligence, and direct perception. (Direct perception is vital to modern ESP theories based on perception of the outer world without the mediation of the sensory organs.) Bertrand Russell was of the opinion that the Platonic position was the seed of the decay that gradually led to the superstitions of the Dark Ages.

Philosophical skepticism, a position of doubt about the possibility of precise knowledge, follows naturally from idealism. If the only things I know for sure are the thoughts within my own mind, then there is no way I can be sure of truths concerning the world outside of my mind. This is especially true if I am perceptive enough to realize how easily and subtly the ideas within my mind change from time to time.

Realists start with the assumption that things exist outside of us independently of our thoughts. Our perceptions of these objects arises from information originating in these outside things and received by our sensory organs through light, sound, and touch. Once inside the body, this information takes the form of electrochemical signals that may be processed by the nervous system. The physical brain then interprets these signals to construct a mental picture of what exists out in the real world. These are the premises of contemporary neuroscientists.

The practice of modern science is, to an overwhelming extent, based on the philosophy of realism. In spite of this, modified forms of idealism have been carried over to remain as powerful influences in modern thought. A number of theoretical physicists, as we shall see later, speak in a remarkably idealistic manner.

Within the past century the development of science and technology has changed our concept of reality, making it qualitatively different from what it used to be. The key to this difference is the growth of instrumentation. Modern scientists do not depend only on human sensory organs to observe the properties of the things out there in the world. They use instruments: cameras, telescopes, microscopes, spectroscopes, cloud chambers, photo detectors, particle detectors of many kinds, etc. We do not expect instruments to have emotional biases. Thus, when several instruments reach the same conclusion about the mass of an electron, or the structure of an atom, or the velocity of a light beam, we have a high degree of confidence concerning the accuracy of this finding.

Anyone who has ever watched a bright track bloom into existence within a cloud chamber has no doubt that something real—a particle—has flashed through the chamber causing the track to form. It is not even necessary for anyone to watch the track. Instruments can scan tracks and analyze data with the aid of computers that perform pattern recognition in the same manner as humans. Thus, instrumentation has transformed both science and philosophy. It reduces the uncertainty about the reality of what we see, and it eliminates the distortion of sensory perception to which we are all prone.

Instrumentation also provides a powerful argument against some forms of idealism currently popular in certain quarters. There are a number of writers who would have us believe that humans actually create reality instead of simply observing it. This idealistic position can be interpreted in two ways: (1) humans create the structure of the universe, and (2) humans create the objects within the universe. What does all this mean?

1. If ideas within the mind are primary, it follows that theories come first and the universe adjusts its behavior accordingly. When a new scientific theory is created within someone's mind and is then veri-

fied, it is, according to this notion, because nature has shaped itself to conform to the new idea. Consider, for example, Dirac's prediction of the existence of positrons. The prediction came first and positrons were then observed, as though the theory caused positrons to spring into existence. Q.E.D.

Unfortunately for this idea, there are too many counter-examples that refute it: theories that were proposed but never verified. What, for example, has become of the ether—the mysterious fluid which was supposed to carry light through space? This was an idea believed by scientists for many decades, only to be replaced by another idea that passed all the tests. Did the ether exist while it was believed, only to vanish from the universe when faith failed? It seems unlikely.

Another argument against the notion that humans create the structure of reality is the fact that many phenomena were observed before there was any theory to explain them. It is a historical fact that electrons were observed before anybody knew what they were, before they were named, and before anybody had a theory about them. A lot of people saw the mysterious green glow emanating from the glass envelope of a vacuum tube when high voltage electricity was applied to the tube's electrodes. Nobody could imagine what was carrying electric current through the airless space, but after J. J. Thomson measured the behavior of this current by applying electric and magnetic fields to the vacuum tube (in 1897) he realized the current was carried by a stream of tiny negatively charged particles. Not until years later did scientists begin to call these particles electrons.

The only way to understand this chain of events is to admit that the electrons existed from the start. Our knowledge of them came later —from interpretation of instrument readings and sensory data. Because of their experiences, physical scientists—especially experimentalists—are overwhelmingly realistic rather than idealistic in their practical, day to day work.

2. Idealism, in the sense that humans create the things they observe, shows up in many interpretations of quantum theory. In this theory the nature of reality at the atomic level has become quite ambiguous. In particular, the problem of finding a meaning for the quantum-mechanical wave packet continues to generate a great amount of controversy. It has also created a grand outpouring of pseudoscientific and pseudo-philosophical writing. Nick Herbert has provided a useful

survey of idealistic ideas embraced by some of the most prominent physicists.[3] Many of these theories imply that an elementary particle does not exist until it is observed, and that the very process of observation brings the particle into existence.

Consider the simple problem of detecting an electron emitted by the cathode of a particle accelerator. While that electron is in flight, I have no way of knowing where it is located and in which direction it is going. Nothing is definite about that electron until it is detected— until it leaves a permanent mark such as a black spot on a photographic film. Not until then can anybody know where the particle is. But does not knowing where the particle is mean that it does not exist? Many scientists speak as though this were the case.

Werner Heisenberg, a dyed-in-the-wool idealist, is quoted as saying, "The idea of an objective real world whose smallest parts exist objectively in the same sense as stones or trees exist, independently of whether or not we observe them . . . is impossible." On another occasion Heisenberg wrote, "The conception of objective reality . . . has thus evaporated . . . into the transparent clarity of a mathematics that represents no longer the behavior of particles but rather our knowledge of this behavior."[4]

Paul Davies writes, "According to Bohr, the fuzzy and nebulous world of the atom only sharpens into concrete reality when an observation is made. In the absence of an observation, the atom is a ghost. It only materializes when you look for it. And you can decide what to look for."[5]

John Gribbin, after a reasonable discussion of the wave packet concept and detection of electrons, states: "Nothing is real unless we look at it, and it ceases to be real as soon as we stop looking."[6]

What are we to make of these comments? Heisenberg, as an idealist, was emphasizing his belief that our knowledge of an elementary particle is based solely on our observations, and that knowledge is contingent on our manner of observing or detecting the particle. The object itself is indeterminate in properties until the act of detection "reduces the wave packet" (Heisenberg's words) and specifies the properties. In a certain sense the particle and the detector make up a single system and cannot be separated from each other in the act of detection. Choosing the detector determines what properties of the particle you are going to detect. Bohr carries the metaphor further, claiming that the particle materializes only when you look for it, implying that it is a conscious human being making the observation. Then Gribbin, consciously or unconsciously, extends the argument to all objects—not just elementary particles: "Nothing is real unless we look at it . . ."

All of these quotations demonstrate philosophical idealism, advancing the concept that nothing exists unless somebody observes it—an extension of the hoary question about whether a falling tree produces a sound if there is nobody in the forest to hear it. This attitude, in fact, was typical of many who adopted the "Copenhagen interpretation of quantum mechanics," which dominated physics for a period of 50 years, which says, in brief, that " 'objective reality has evaporated', and that quantum mechanics does not represent particles, but rather our knowledge, our observations, or our consciousness, of particles."[7]

Understanding quantum theory inevitably requires interpretation. The question is whether the interpreters are going to be realistic or idealistic in their philosophy. It is not only a matter of taste. Only realistic interpretations can pass the tests of empiricism necessary before a theory can be considered "scientific." An empirical theory is a testable theory, and in order to test a theory it must be possible to describe in a precise manner the operations necessary to perform these tests.

When we interpret quantum theory in a realistic way, we find the following: Quantum theory simply states that certain properties of particles (such as position or velocity) are not well specified until the particle is detected. When an electron passes through a small aperture we do not know exactly where it is, because the aperture has a finite width. And we do not know where it is going until it actually comes to rest in a detector. A number of identical electrons passing through the same aperture with the same velocity may end up in widely different locations. This is the meaning of Heisenberg's uncertainty principle.

It is the word "detector" that causes much of the confusion concerning the existence of the electron. Many writers and physicists metaphorically use the word "observer" in place of "detector." The reader is then tempted to infer that a conscious observer is required. Nothing could be further from the truth. Detection does not require the presence of a human observer. Quantum theory says nothing about conscious observers, human or otherwise. (Even Heisenberg finally admitted that "it does not matter whether the observer is an apparatus or a human being."[8])

Any process by which a particle interacts with its environment to produce a permanent change in that environment qualifies as an act of detection. It may be nothing more than the darkening of a silver granule on a photographic film. It may be a sparkle in a scintillation counter, or a track in a bubble chamber. A human may not see the black silver spot on the film until a thousand years after the spot was created. Does this mean the particle responsible for the silver grain did not exist until the human finally decided to look at the film? Are we to believe that a particle detected in a scintillation counter does not exist until somebody reads the numbers recorded by the computer to which the counter feeds the electrical pulses? This kind of interpretation is an overheated parody of what quantum theory really says. But that is what the idealistic writers would have us believe.

A more realistic interpretation is this: A fundamental particle is represented mathematically by a wave function. The wave function does nothing more than allow you to calculate the probability of finding the particle in a given state or in a given location. Until it is detected by a non-reversible interaction with something in its environment, the state of the particle is not well-specified. It may be found in any number of locations; it may be in any number of energy states.

Most significant, the machinery set up for detecting the particle determines what property is going to be specified. When you place a photographic film in the path of the particle and the particle leaves a black spot, then its position is determined. When you place a velocity analyzer (a device with electric and magnetic fields) in the path of the particle and the particle passes through (and is detected), then its velocity is determined. But, says quantum mechanics, you cannot determine both the position and velocity of the particle simultaneously with total precision.

Also significant—and often ignored in these discussions—is the fact that it is not necessary for a human to arrange the detection apparatus. A million years ago a cosmic ray particle may have plowed into a rock put in place by natural forces and untouched by human hands. Did the particle exist a million years ago, or did its existence have to wait for a human to come along and see the track left behind by the particle? If you reply "It had to wait," then we have come to a *reductio ad absurdum*.

Easily lost in all this philosophical verbiage is the fact that humans do not experience elementary particles directly. It is true that the retina of my eye can respond to the impact of just a few photons, sending a signal down the optic nerve. However, I am not aware of that response until a very complex set of amplifications and signal processing has taken place in my nervous system. It is a long way between absorption of a photon in the retina and knowledge that something has been seen. Are we to say that the photons do not exist until I am conscious of their existence? If the answer is yes, then we imply reverse causality: knowledge of the observation causes the existence of the photon at an earlier point in time. Another *reductio ad absurdum.*

The topic of "observation" has a way of inspiring serious writers to make grievous semantic lapses. Consider the observation of a star many light-years away. Is the star non-existent until I observe it? Of course not. For it is not the star itself that impinges on my retina, but rather a number of photons emitted from the star—photons which may have traveled for thousands or millions of years before entering my eye. *It is the photons that I detect, not the star.*

The word "observation" is seen to be a high-level abstraction that may involve a large number of steps leading from the observed to the observer: the emission of photons by the observed star, transmission of photons through space, absorption of these photons in the retina, transmission of electrochemical signals through numerous synapses until they reach the brain. And then what? A good many writers speak of the information reaching the "mind," which then becomes conscious of the originating event. But what if we substituted a video camera and a computer memory for the human observer? Would it make any difference to the process of observation? As far as physics is concerned, there is no difference between storing the signal from the star in the human brain and storing the signal in a computer memory. If this is so, then why should *human* observation make a difference? Again, we read idealism in the writings of those who say that it does. Any talk of humans observing stars directly is merely pre-scientific jargon without logical meaning.

Much commotion in the world of quantum theory was caused by the famous theorem of J. S. Bell,[9] who investigated the problem of two photons that are emitted simultaneously from an atom and then travel in opposite directions away from the atom. Since these two photons are described by a single wave packet, if something affects one photon, it must instantly have a corresponding effect on the other photon, even though the two photons have become separated by a great distance. Bell's theorem gives us a formula for calculating observable effects resulting from this phenomenon.

These effects were demonstrated experimentally (in 1982) by a group headed by Alain Aspect.[10] The apparent instantaneous transmission of information from one photon to the other attracted the attention of many writers who tried to use this phenomenon to justify a belief in telepathy and other aspects of mystical experience. The possibility of faster-than-light communication was immediately denied by both Aspect and Bell, the two physicists closest to the problem. (More about this under Myth 3.)

Bell, the leading researcher in the theory of quantum reality, is himself scornful of the notion that a human observer is necessary for the existence of an observed object. He says: "The only 'observer' which is essential in orthodox practical quantum theory is the inanimate apparatus which amplifies microscopic events to macroscopic consequences. Of course this apparatus, in laboratory experiments, is chosen and adjusted by the experimenters. In this sense the outcomes of experiments are indeed dependent on the mental processes of the experimenters! But once the apparatus is in place, and functioning untouched, it is a matter of complete indifference . . . according to ordinary quantum mechanics . . . whether the experimenters stay around to watch, or delegate such 'observing' to computers."[11]

Even Albert Einstein, while fundamentally a realist, was occasionally baffled by the problems encountered while trying to understand the world. During his later years he wrote, perhaps in a playful mood, "He [the scientist] must appear to the systematic epistemologist as a type of unscrupulous opportunist: he appears as realist in so far as he seeks to describe a world independent of the acts of perception; an idealist in so far as he looks upon the concepts and theories as the free inventions of the human spirit (not logically derivable from what is empirically given); as positivist in so far as he considers his concepts and theories justified only to the extent to which they furnish a logical representation of relations among sensory experiences."[12]

If we want science to be a successful enterprise, we must have a way of deciding whether one theory is more valid than other theories. In this modern period, pragmatism is the final judge. Does the theory work? Does it pass the experimental tests? Above all, is the theory empirical; can it be falsified? And has it been verified by properly designed experiments?

Many philosophers of science believe that the same criteria apply when deciding between idealism and realism as universal models. Our goal is a pragmatic test that allows us to choose the better philosophy— better in the sense that it gives better results than competing methodologies. In principle such a test exists.

First, we should recognize that idealism (in its most extreme form) cannot be falsified by any experiment or observation, because if reality is created by your own mind, then any results obtained by an experiment can be interpreted as something originating in that mind. No other conclusion can be drawn. As Karl Popper has pointed out, any theoretical system that cannot be falsified in principle is not an empirical system of knowledge.[13] Consequently, idealism cannot be considered an empirical system of knowledge, and so falls outside the domain of science.

Secondly, we shall ask whether an idealistic theory has any explanatory value. In idealist philosophy the concept of "mind" as an entity apart from the body is fundamental. Yet there is never an attempt to explain the "mind" in empirical terms, no possibility of experimenting directly with "mind" or measuring its properties, and above all there is no attempt to explain how "mind" interacts with body. To the idealist the mind is an ethereal something, but what it is, nobody knows.

The only viable alternative to idealism is realism, in which "mind" is the end result of the operation of the nervous system. It is the totality of the patterns created by the signals stored in the molecules of the brain. The only hypothesis possible to realists is that mind is a high-level abstraction at the top of a continuum of abstractions in which fundamental particles are at the bottom. Particles join to form atoms, atoms form molecules, molecules form cells, cells arrange themselves into nervous systems, and nervous systems are the conduits and the processors of signals and information. Out of these signals arise language, thought, creativity, and awareness. To prove the truth of these statements is the agenda of modern neuroscience and information theory.

This is not to deny the strangeness of quantum physics. I am merely agreeing with J. S. Bell that "mind" does not have anything to do with things happening outside of the body, and that the universe would go along as it always has even if no human beings existed. Indeed, since the universe existed for many billions of years prior to the existence of humans, extreme idealists who claim that consciousness of the world is necessary for its existence are behaving just like creationists. To them consciousness and existence began simultaneously a relatively short time ago.

The physical study of thought processes, the province of neuroscientists and information-scientists, provides us with the ultimate empirical test of realism as a philosophy. We start by asking what the necessary and sufficient conditions are for realism to be a correct theory.

Realism is a sufficient theory if we can use physical principles to explain the mechanism by which consciousness arises from the workings of the brain. The theory is strengthened if we can design a computer that creates concepts and theories and is aware of its own existence. Realistic neuroscience has had some success in describing the operation of simple parts of the nervous system. Computers designed according to the principles of neural networks are beginning to simulate the operation of the human mind. At present it is far too early to predict how far we can go in this direction, because the study has just begun.

Realism is a necessary theory if it is the only theory that works. If idealism is unable to explain the mechanism of consciousness, then realism is necessary, for it would then be the only theory. As we have seen, idealistic and mystical theories have made no progress in explaining how the mind works, whereas the direction of positive and fruitful research results points in the direction of realism.

Pragmatism is thus the decisive factor. The bottom line becomes: Does the theory work? Does it enable us to design devices that work? Most important: does it enable us to understand how the universe works and how all the occurrences in the universe are caused? At present realism works successfully in the domain of inanimate objects. Realism enables us to explain how things operate in the world of the non-living and the non-conscious. It does not as yet explain how living is different from non-living, although it is getting very close. We are not yet prepared to say how consciousness is different from non-consciousness, except that we think all conscious things have nervous systems and non-conscious things do not.

On the other hand, idealism is capable of no explanations at all— it yields nothing but fantasies.

For example, idealism says nothing about why ten different observers in different parts of the world measure the speed of light to be the same. If the light beam exists only as a construct in my mind, then how does an experimenter in Moscow always get the same result that I do in, say, Princeton. Idealists may issue vague pronouncements about the way my personal reality is connected to the reality existing within the minds of all others. But this is the kind of unverifiable and unfalsifiable "explanation" which demonstrates only that idealism, in reality, explains nothing.

Science works best when it is based on reality. For this reason, the balance of this book exhibits a strictly realistic point of view. The development of science during the past three centuries—both in its rate and in its direction—demonstrates that this is the most fruitful approach.

At this point some of you are going to raise an obvious objection. If idealists such as Heisenberg and Bohr were responsible for the development of quantum theory, how can I claim that realism is the most useful approach to science?

The answer is simple. The creators of quantum theory were products of the nineteenth century, and they were steeped in nineteenth-century philosophy. In this philosophy there were two schools of thought: those who adhered to the theories of idealism, and those who believed in the methods of positivism. Idealism pervaded their private thoughts and their personal interpretations of quantum theory. Positivism provided the guiding principle that in practice quantum theory must provide connections between observables, and that observables are the only entities allowed in the discussion.

In their technical writings, attention was concentrated on realistic science: on the development of equations that described the structure of atoms and permitted calculation of the wavelengths of light emitted from these atoms. Insofar as their work was realistic, it was successful. The wavelengths of light could be measured; the measurements agreed with the predictions of the theory. But when they tried to interpret the theory and made idealistic statements to the effect that particles do not exist unless they are observed, then they descended into error. These ideas no longer carry much weight outside of pop-science writing.

The moral is that the private opinions of good scientists need not interfere with their work.

Reality, Fantasy, and Myth

The physicist's recipe for the universe is fundamentally simple: take three different types of particles (quarks, leptons, and bosons), stir them into motion with the four kinds of forces by which they interact, and let the mixture cook. The universe then proceeds to do what it has to do. The fact that human beings arise out of this inanimate stew is still somewhat mystifying. However, the fact that we do not as yet understand how humans arise from particles should not discourage us, and should not lead us to assume that such understanding is impossible. The study has just begun.

To most humans, a universe consisting of particles banging about and doing what they have to do seems cold, barren, and without meaning. "Meaning," however, is not something that floats in space, permeating the universe like a nebulous, mystical cloud. A comic-strip guru who sits on a mountain top and claims to know the true "meaning of existence" is perpetuating a myth. "Meaning" arises out of the working of the human mind, and therefore exists only in the human mind. The meaning of existence is whatever you want to make of it. (This comment applies also to any sentient being who might inhabit any other planet.)

Scientists do not seek universal meanings in the manner of the mountain-top guru. This kind of meaning is not within the domain of science. Scientists merely seek knowledge of the world that exists, and they seek explanations of why things happen the way they do. But the "why" of science is not the same as the "why" of the victim of misfortune agonizing over "Why me?" To explain why things happen the way they do, the scientist looks for causes in terms of the rules that determine how particles move and interact, how chemicals react, and how organisms respond to their inner and outer environments. The "why" of science deals with causes, not with motives or purposes. To the scientist, the thing that happens to one person would happen to anybody else in the same shoes. It's not a personal matter.

In nature things do only what they are allowed to do. Never do they do what they are not allowed to do. Humans, possessing consciousness, perceive the regularities of nature with the aid of their instruments and codify these regularities into rules that they call the laws of nature. These laws are then arranged into an intellectual structure called science. Nature, not having consciousness, knows nothing about laws in the human sense, and cares nothing about the words or equations that human beings use to describe them. Nature simply possesses a specific structure, and this structure imposes upon the objects within nature certain necessities. As a result of these necessities, some actions are allowed to occur and some actions are not.

It's like a cosmic chess game. A bishop can move only along a diagonal. It is not allowed to do anything else under the rules of the game. Similarly, the earth is required by the geometry of spacetime to move in an elliptical orbit around the sun. With the amount of energy it happens to have, it cannot do anything else.

None of this should be confused with the theological concept of "natural law." When conservative clerics issue judgments such as "Homosexuals perform acts against the laws of nature," what they mean is that they believe homosexual activity is forbidden by God, because in their lexicon, the laws of nature are laws established by God. Physics, on the other hand, takes as its starting point observations of nature itself: its structure and its dynamics. Anything that happens happens because the laws of nature allow it to happen. Anything that is forbidden by the laws of nature simply does not happen. From this point of view, an action performed by over ten percent of the population is clearly not forbidden. This is not to deny that unpleasant consequences may result if care is not taken.

The natural law of theology and the man-made laws created by legislatures are seen to be quite different in operation from the natural laws of physics. Theological and legislative statutes say: if you do what is forbidden (and get caught) you will be punished. The physical laws of nature say: you simply cannot do that which is forbidden.

The most basic laws of nature control the destiny of objects at the atomic level and below: they determine how fundamental particles interact with each other, and how atoms combine with each other. The atoms that make up human beings follow the same laws. We do not have one kind of atom for animate creatures and another kind for inanimate objects. That sort of idea became obsolete in 1828, when Friedrich Wöhler synthesized the compound urea without making use of animal urine.

Because humans are considerably more complicated than atoms, the elementary laws that govern the activities of individual atoms provide us with very little help in understanding how humans behave, or in predicting what humans are going to do. This fact leads to some confusion concerning the validity of the philosophical concept of "reductionism"—the idea that all phenomena may be reduced to atomic physics. Because of the complexity of human behavior, reductionism would appear to be a simpleminded theory, a fact seized upon gleefully by those who insist that a human being is more than simply a collection of atoms and molecules. But there is nothing novel about the fact that many particles in large, interactive structures can do things that single particles cannot do. Turbulent motion requires the concerted action of numerous particles in a fluid; signal processing requires solids arranged in electronic circuits. Anti-reductionists, however, appear to operate on the belief that the workings of living (especially human) beings are of such a special nature that extra-scientific, mystical, or supernatural elements are required to explain them. (As though mysticism offers any kind of rational explanation.)

On the other hand, reductionism acquires a more encouraging aspect if it is applied in reverse. Instead of hopelessly trying to use the laws of physics to predict what humans can or will do, what happens if we use these laws to predict what humans *cannot* do? This turns out to be quite a profitable tactic, for if a certain action is forbidden to a single atom by one of the fundamental laws of nature, this action must be equally forbidden to any assemblage of atoms.

At this point you might wonder if there is a contradiction here. On the one hand we say that an action forbidden to a single particle is forbidden to any number of particles. On the other hand we say that a single particle does not engage in turbulent motion, while many particles do. It is important to note carefully the difference between these two statements. Certain kinds of activities *require* the cooperation of a large number of particles and cannot take place with just a few particles. Turbulent motion of fluids is one such activity, and is described by a *top-down* rule that stems from the way turbulent motion is instigated. The action starts with an equation describing the collective motion of many particles, and simply does not apply to a single particle. But the statement that energy cannot be created or destroyed is a *bottom-up* rule that starts with a single particle or a few particles, and continues to apply when the number of particles is increased to ten or to ten trillion atoms. The fundamental laws of physics are bottom-up rules.

As I suggested earlier, we may divide the laws of physics into two classes. *Laws of permission* are laws which specify what kind of actions are allowed in a given situation. *Laws of denial* are laws which specify what kind of actions may not take place.[14] Laws of permission say that particles may engage in turbulent behavior if there are enough of them, and if the environment encourages turbulent behavior. Conservation of energy, on the other hand, says that the particles within a closed system can do nothing that changes the amount of energy in the system, regardless of how the particles are arranged, or how many or how few there are.

The concept of permission versus denial is an idea not commonly stressed in elementary physics courses, yet it is implied within every physical theory and every equation describing the behavior of physical objects. We will pursue this topic in more detail in Myth 2, but in the meantime we will use the concept to forge a general principle: *while we usually cannot make precise predictions about what people will do in a given situation, we can state, with great confidence, what they cannot do.* Specifically, they are not allowed to do anything that implies a violation of one of the laws of denial.

Predicting what people are not allowed to do might seem a trivial occupation. Not so. Some of the most monumental scientific predictions are negative in nature. We have already seen that no matter how many atoms are involved in an action, they are not allowed to do anything that requires the creation of energy out of nothing. This basic rule has the most far reaching consequences.

For example, it is this principle that allows patent examiners to disregard claims for perpetual motion machines a priori. They don't have to squander taxpayers' money examining every proposed device purporting to create energy out of nothing. These contraptions simply never have worked, do not work, and cannot work.

Can we make similar negative predictions about human beings? Certainly. I can predict, for example, that none of you is going to jump to the moon without the aid of mechanical devices. Is this prediction too obvious to be taken seriously? Perhaps. But when I predict that nobody in the world will ever levitate himself above the ground merely by thinking about it, I am disputing the claims of numerous cult members who say they can do it. And when I predict that nobody will ever send telepathic messages from the earth to the moon, or from the moon to the earth, I am contradicting the claim of an ex-astronaut who says he has done this.[15]

This is a most profound and powerful generalization: the same laws of nature that limit what atoms can do, also limit the actions of human beings. Despite all their delusions of grandeur, technological hubris, and exponential increase in knowledge, human beings are still bound by the structure of the universe to behave within the strict limits of physical law.

It is a conceit of humans that they, with their modern scientific tools, have gained control over nature. However, nature is not that easy to control. We can never force things in nature to do anything that they do not naturally do. When my fingers tap the keys on this computer, I get a sensation of willing something to happen. But what do I see when I look inside my finger? The more closely I examine what happened within the nerve fiber and muscle tissue, the less evidence I find of volition. I find nothing but electrons and ions moving under the action of impersonal electromagnetic forces, following the rules of quantum mechanics.

Free will seems to vanish under the searchlight of analysis. Is there nothing left but brute determinism? Probably not. If we try to reduce the choice to that between free will on the one hand, and determinism on the other, we propagate another of those myths about the nature of science that permeates public discourse. Nature is too complex to be described in terms of such simple choices. It is becoming evident that neither free will nor determinism can all by themselves describe the nature of consciousness. Even in strictly determined systems the presence of feedback mechanisms produces chaotic results that defy our ability to make predictions of future behavior. A system that is completely deterministic on a microscopic level may behave in an apparently non-deterministic manner when all the different inputs and memories and feedback mechanisms are added to the equation.

The direction of development in the biological sciences leads us to believe that we can eventually reach an understanding of the mechanisms of consciousness and of the meaning of meaning. Seeing how far we have come in a relatively short time encourages us in our hope that the problem can be solved. The more we learn about the working of the nervous system, the more we realize that the progress of the last few decades has been made possible by the development of realistic concepts and elimination of idealistic myths of the past.[16]

It is hard for many people to accept the idea that "mind" can arise out of matter, and that consciousness can be explained on a physical/mathematical basis. But the alternative theories—based on undetectable psychic forces—are even less satisfactory as explanations.

I state these opinions from the viewpoint of one who has watched physics change and grow during the past half-century, and has watched with some dismay the fantasies sold to the public not only as science fiction but under the guise of hard science. Books on science written for the layman are most successful when they appeal to the sense of wonder and mystery aroused by a contemplation of the universe at large. This is all to the good. However, sales are even better when the writing hints at a connection between the mysteries of physics and the seductions of mysticism. (One result is that some physics books end up on shelves labeled occultism, and vice versa.)

Yet the urge to go in that direction can be resisted. Stephen Hawking, one of the great physicists of the century, and author of a best-selling book on cosmology, has been quoted as saying, "There is no sharp boundary [between mathematicians and mystics], just a gradual descent into wooliness."[17] In physics, wooliness of thinking makes bad science. Mysticism and science are diametrically opposed modes of thought, since mysticism implies direct perception of truth, and science requires verification of ideas through the senses and through instrumentation. (Nevertheless, mysticism as a human phenomenon does exist, and can be discussed within a framework of science, as Paul Davies demonstrates.[18])

The tenuousness of communication between scientists and non-scientists is graphically displayed whenever a scientist gets into a discussion with a non-scientist about any one of a number of topics. Try arguing with a devout science fiction reader who believes that some day we will be able to go to the distant stars faster than the speed of light. "Nothing is known for sure," he is bound to say sooner or later. "Some day an advanced civilization will find a way to travel faster than light. Anything is possible." In this one argument, three myths about science have been put on display. So pervasive are these myths that quite a number of scientists believe them to be true, even though a moment of thought would persuade them that their primary function is propaganda rather than meaning.

The myths about science date from a period when science was just beginning, and when it was not clear how to separate reality from fantasy. This was a time when most scientists refused to believe in the existence of atoms, but did believe in unrealistic things called "ether" and "caloric." A hundred years of research has caused an enormous change in our understanding of nature, and we are now much more certain about what we know and what we don't know. Yet the myths continue, and they are the cause of a number of social problems.

As we put the myths about science under the microscope, we will see how they are, in common with most myths, a mixture of half-truths, wishful thinking, and outright fantasy. We will also see how much harm these myths can cause.

NOTES

1. B. Russell, *A History of Western Philosophy* (New York: Simon & Schuster, 1945), p. 73.

2. Ibid., p. 121.

3. N. Herbert, *Quantum Reality* (Garden City: Anchor Press/Doubleday, 1985), chap. 2.

4. W. Heisenberg, "The Representation of Nature in Contemporary Physics," *Daedalus* 87 (1958): p. 95., quoted by K. R. Popper in ref. 7, p. 42.

5. P. Davies, *God and the New Physics* (New York: Simon & Schuster, 1983), p. 103.

6. J. Gribbin, *In Search of Schroedinger's Cat* (New York: Bantam Books, 1984), p. 173.

7. K. R. Popper, "Quantum Mechanics Without 'The Observer'," in *Quantum Theory and Reality,* ed. M. Bunge (New York: Springer-Verlag, 1967), p. 7.

8. M. Bunge, *Quantum Theory and Reality* (New York: Springer-Verlag, 1967), p. 4.

9. J. S. Bell, "On the Einstein-Podolsky-Rosen Paradox," in *Speakable and Unspeakable in Quantum Mechanics* (Cambridge: Cambridge University Press, 1987), pp. 14–21.

10. A. Aspect, J. Dalibard, and G. Roger, "Experimental Test of Bell's Inequalities Using Time-Varying Analyzers," *Physical Review Letters* (20 Dec. 1982): p. 1804.

11. J. S. Bell, "On the Einstein-Podolsky-Rosen Paradox," p. 170.

12. A. Einstein, "Reply to Criticism," in *Albert Einstein: Philosopher-Scientist,* ed. P. A. Schilpp (La Salle, Ill.: Open court, 3rd ed., 1969), p. 684.

13. M. A. Rothman, *A Physicist's Guide to Skepticism* (Buffalo, N.Y.: Prometheus Books, 1988), p. 164.

14. Ibid., chap. 5.

15. E. D. Mitchell, "An ESP Test from Apollo 14," *Journal of Parapsychology* 35 (1971): p. 89.

16. See R. L. Gregory, *Mind in Science* (Cambridge: Cambridge University Press, 1981); J. Winston, *Brain and Psyche* (Garden City, New York: Anchor Press/Doubleday, 1985); and F. I. Dretske, *Knowledge and the Flow of Information* (Cambridge: The MIT Press, 1981).

17. M. W. Browne, "Mystics and Science: Hawking's Views," *New York Times* (19 April 1988): p. C5.

18. P. Davies, *God and the New Physics*.

MYTH 2

"Nothing is known for sure."

Confident Knowledge

The myth that "Nothing is known for sure" is diverse in its uses. It may convey either an air of false modesty, or of dogmatic skepticism. Consequently it may be used in argument by a wide variety of people who range from the legitimate scientist to the paradoxer defending a pet pseudoscientific theory. It may be used on the one hand to defend the validity of pragmatic knowledge, or on the other hand to assert that absolute knowledge is impossible. It is an all-purpose conversation-stopper.

In offensive debate it is always used to cast doubt on the certitude of the opposition. The fundamentalist espousing "creation science" tells the geologist: "You don't know anything for sure. The rates of radioactive decay may have changed during the last ten thousand years, and if you are not certain about your radioactive dating measurements, you can't know how old your fossils are. Therefore your whole theory of evolution is no good." What can the geologist say in response?

As a defense, the argument that "Nothing is known for sure" serves to create an atmosphere of sweet reasonableness. If you tell the hopeful inventor of a perpetual motion device that his machine cannot possibly work, his answer is likely to be: "You don't know anything for sure. You've never tested this particular kind of machine, so you can't be sure that perpetual motion is absolutely impossible."

Lost in this controversy is the fact that this argument is a two-edged sword. If nothing is known for sure, then how can the creationist be more certain than the evolutionist about the truth of his ideas? How can the perpetual motion fanatic be more sure than the skeptical physicist about the value of his invention? The fact of the matter is that a feeling of confidence may arise from factors that have nothing to do with the merits of the theory. The person with the fewest facts usually is the one most certain of the truth. The scientist, backing up his ideas with a world of empirical evidence, may be more modest in his claims of total certainty.

This state of affairs was recognized by Jacob Bronowski, a mathematician, poet, writer, and a superb interpreter of science to the public. Prior to his death in 1974, Bronowski had appeared as narrator in a television series that dealt with the development of humankind's ideas about nature. The text for this series became a book titled *The Ascent of Man*.[1] In a chapter titled "Knowledge or Certainty," Bronowski makes a striking statement: "There is no absolute knowledge. And those who claim it, whether they are scientists or dogmatists, open the door to tragedy. All information is imperfect. We have to treat it with humility."

The first response to this denial of absolute knowledge is incredulity. After hundreds of years of scientific effort, do we know nothing for certain? But as we read further into Bronowski's text we learn that he was not denying knowledge, but was instead comparing empirical knowledge based on less-than-perfect physical evidence with the certainty of absolute belief based on no facts at all. He was comparing pragmatism with dogmatism. In particular, he was comparing the Heisenberg uncertainty principle with the pseudoscientific racial theories of the Nazis—theories whose effects destroyed the scholarly tradition of Germany while at the same time destroying the lives of millions of Europeans.

Werner Heisenberg, one of the founders of quantum theory, was a theoretical physicist at the University of Göttingen, in Germany. The uncertainty principle that bears his name is a rule based on laboratory observations of the particles which make up all matter: electrons, protons, neutrons, photons, etc. These observations force us to conclude that the object we call a particle is not what we would like to think of as a particle; it is not a hard ball with a definite size, position, and velocity. Its properties have a curious fuzziness, or uncertainty, to them. If we know the particle's position exactly, then we cannot know how fast it is going. If we know exactly how fast it is going, we cannot know where it is located. If we know its position approximately, then we can measure its velocity approximately. The uncertainty in the position is related to the uncertainty in the velocity by a simple formula that is the core of the Heisenberg uncertainty principle and the heart of quantum mechanics.

Since we cannot know everything there is to know about an object with perfect precision, our knowledge about it can never be total. This imprecision applies to all kinds of measurements. No matter how good the instruments used to measure a position, a velocity, a voltage, or a temperature, the results of these measurements can never be absolutely perfect. There is always some error, ultimately caused by the uncertainties in the motions of the particles within the measuring device, as well as the uncertainties in the state of the things being measured.

But, granting that all measurements are uncertain, does it follow that all knowledge is imperfect, or that we know nothing for sure?

To a certain extent, the answer to this question depends on the mental state of the person supplying the answer. There are some people of an idealist bent who are so modest (or so fearful), that they are not even sure of their own existence. They are like Tweedledee and Tweedledum, who thought that they—and Alice—were just things existing in the Red King's dream, and wondered what would become of them if the King left off dreaming about them. Others, on the other hand, are so puffed up with self-esteem that they claim absolute certainty about all opinions that fill their minds.

If we are not to be totally paralyzed in action, we must choose some kind of basic philosophy, some model of the universe. As described under Myth One, in this book I choose a model that favors realism over idealism. The reason for this choice is pragmatism: realism has enabled us to explain a great deal about what we see in the universe, while idealism explains nothing. Whenever it does attempt explanations, its logic is based on the assumption of entities whose existence is never proven. In addition, realism has predictive powers. Logic based on realism enables us to predict the motions of planets and space vehicles, while idealism enables us to predict nothing. (Often idealists predict real events, but when they do they are actually using realistic theories. They just choose to interpret their theories in an idealistic way.)

When experimental physicists build particle accelerators, it is be-cause they want to create real particles. When they set up arrays of particle detectors, it is because they believe there are particles out there to detect. These particles, when they join together in multitudes to form rocks and animals and planets, create real objects. If we do not agree on that premise, then knowledge cannot exist. Reality testing enables scientists all over the world to agree on the validity and accuracy of their theories.

Having settled on reality as the most fruitful road to knowledge, we now investigate how firm this knowledge is. Is there actually such a thing as absolute knowledge? When we try to answer this question, we discover that there are several different kinds of knowledge, and each provides a different answer to the question.

1. Some knowledge is based solely upon definition of terms, and therefore is perfectly precise within the context of its definitions. All of mathematics is of this nature. In mathematics two plus two equals precisely four in the domain of real numbers because this is the way numbers—and the operation of addition—are defined. You can make other definitions, but then you have a different kind of mathematics— vector addition, for example.

When we relate mathematics to the physical world, other rules may apply. In physics, two kilograms plus two kilograms are less than four kilograms. When we put two masses together, the combined mass is less than the sum of the individual masses. The reason we never notice this when we deal with kilogram masses is that the difference is ex-ceedingly small. But when we put two atomic nuclei together, the loss of mass is very important, and is responsible for fusion energy. In the real world we deal with the way objects actually behave, and defini-tions must conform to this behavior if they are to have any significance within physics.

2. Certain types of observations in the physical world are abso-lutely certain, because the uncertainty principle does not apply to them. For example: one thing we know for sure is that two electrons always push each other apart. This is the way negative electric charges are supposed to behave. ("Like charges always repel each other" is the rule we learn in school.) A skeptic might raise the question: How do we know that two electrons always repel each other? Perhaps two rare electrons might occasionally attract each other. If this event were very

unusual we might never notice it. Here is where definition of terms comes into play. A particle such as an electron is defined in terms of the totality of its properties. An electron is defined as a particle with a certain mass, a certain amount of electric charge, a certain amount of spin, etc. *And the sign of its electric charge is by definition always negative.* If a particle turns up that is just like an electron, except that its charge is positive, then we call it something else.

As a matter of fact, there do exist particles exactly like electrons, but with positive electric charges. But we don't call them electrons—we call them positrons. And while they are closely related to electrons (a positron is the antiparticle of an electron) they are not the same.

The upshot of all this is that we do indeed see particles that look just like electrons and which attract each other, just as the skeptic claimed. But one of those attracting pairs is not an electron. Two electrons always repel each other. This is qualitative knowledge and so does not depend on exact measurements.

Putting it another way, the sign of the charge (positive or negative) of an electron is a discrete (as opposed to a continuous) quantity, and therefore is not subject to the Heisenberg uncertainty principle. As a result, knowledge concerning the sign of a particle charge may be absolutely precise. It depends only on definition of terms and the observation that the things defined actually exist.

3. There is a class of statements that are accepted as completely true because the measurements that verify these statements are far more precise than needed for this verification. Nobody would doubt that the population of the earth is greater than one million, because this number is far outside the limit of error of even the wildest estimate. For the same reason we can be absolutely sure of being right when we say that the sun is closer to the center of the solar system than the earth. All astronomers will agree to the absolute truth of that bit of knowledge. However, there was a time when you could have been burned for saying that the earth goes around the sun, which amounts to the same thing.

A subclass consists of negative statements, which say that something is *not* something else. For example, we know that the earth is not flat. That is to say, the surface of the earth is not a Euclidean plane surface. Nowadays, with modern instruments to measure the shape of the earth and to show its appearance (a satellite camera, for example), this assertion is a triviality. However, there was a time when arguments over the earth's shape caused considerable excitement.

In the same vein we know with certainty that the earth is not a star, that a horse is not a fish, and that light does not consist of vibrations in a fluid called the ether. The first statement is absolutely true because even though the gradation between planets and stars may be continuous, the earth is so small, and its temperature so low, that it could not possibly be mistaken for a star. The second statement is perfectly true because the differences between a horse and a fish are so great that there is no doubt as to which category any given horse belongs to.

The third statement is also true because it is based on physical measurements that are more precise than necessary to prove the statement. When Michelson and Morley did their famous "ether drift" experiment in the 1880s, they were trying to detect the variations in the speed of light to be expected if the earth was plowing through the ether as it made its way around the sun. The earth's speed around the sun is about 30 km/sec., and the experiment was able to detect an ether drift one-fortieth of that amount. Modern experiments can now detect an ether drift less than one-thousandth of the earth's orbital velocity. We see that these measurements are far more precise than needed to show that the classical ether theory does not work as well as Maxwell's electromagnetic theory combined with Einstein's special theory of relativity. Therefore, we can accept the disproof of the ether theory as perfect knowledge.

Comparisons may also have very high reliability. When we say that Einstein's special theory of relativity describes the universe better than Newton's laws of motion, this is a true statement, for the measurements are far better than necessary to support the comparison. This is not to say that Einstein's special theory is the best theory possible. In fact, Einstein himself made a better theory: the general theory of relativity.

4. Uncertainty must be given its proper respect when we deal with the results of quantitative measurements. Every measurement has some uncertainty to it, either (a) because of fluctuations in the thing being measured, or (b) because of molecular motions within the measuring device. Every physicist is trained to be aware of these uncertainties. An entire branch of physics, usually represented by a single course in graduate school, is devoted to the theory of measurement error. Every physicist (and hopefully every other kind of scientist) is trained to

add error bars to the points on his graphs. These are the little vertical marks that indicate the probable range of error for the measurement. The range of error is represented numerically by a plus-minus sign, for example: 75.2 cm ± 0.5 cm. What this means is that our instrument has measured a length of 75.2 cm, but due to errors of one kind or another, there is a likelihood that the true length is as low as 74.7 cm or as high as 75.7 cm.

Knowledge of probable error is an integral part of our knowledge about the measurement itself. Estimating errors is as important as performing the measurement.

What is not generally realized is how the precision of modern physical measurements puts a new perspective on the idea that "all knowledge is uncertain." Measurements may be so precise that for all *practical* purposes they are completely certain. Instead of worrying about "absolute knowledge," which in many cases may be an unattainable ideal, it is best to assume a pragmatic position and to ask whether our measurements are sufficiently precise to answer questions that might be encountered in the real world.

These arguments explain why I consider the statement "Nothing is known for sure" to be a myth. Some things we do know for sure. Still other things we know well enough. Precision of knowledge is much too complex a topic to be described by a simple saying.

How Precise Can We Get?

One example of exactitude in science demonstrates the power of modern instrumental methods. It shows how precisely the law of conservation of energy may now be verified, and reveals how closely we may come to the ideal of absolute knowledge even when faced with the uncertainties of physical measurement. It also shows us how modern science has altered the answer to the fundamental problem of inductive inference—a problem that has vexed philosophers for centuries: *how do we arrive at a general law of nature that is true for an infinite number of cases when we can only perform a limited number of experiments?* Recent developments in physics have provided new and powerful ways of thinking about this puzzle.

The classical statement of conservation of energy is simple: in any closed system the total amount of energy can never change. That is, no reaction can create energy from nothing, or can make energy disappear into nothingness. The only exception to this lies within the Heisenberg uncertainty principle. When reactions take place during very small periods of time, there is some uncertainty in the energy involved. The smaller the time frame, the greater the uncertainty. But for events taking place in our ordinary time framework, the uncertainties become exceedingly small. How small we shall soon see.

An alternative statement of conservation of energy is the first law of thermodynamics, often quoted to prove the futility of perpetual motion machines. It must be understood that the first law of thermodynamics is not a separate law; it is nothing more than conservation of energy as applied to heat machines. However, conservation of energy is a more general law and relates to all forms of energy, not only heat. It is a law more fundamental than the first law of thermodynamics, since it applies to the elementary particles, and to all objects composed of those particles. It is a law governing the actions of all things.

Here are a few examples showing some obvious and some not-so-obvious consequences of the law:

1. No machine can start or indefinitely maintain any kind of motion that overcomes friction or does work unless it burns fuel or receives energy from an outside source. Whenever a moving machine stops moving, its energy must be converted into heat or some other form of energy. This statement of the law forbids the operation of perpetual motion machines.

2. You cannot transmit information from a transmitter to a receiver without sending energy through the intervening space. The intensity of this broadcast energy always decreases as it spreads out through a greater volume of space. As we shall see in the next chapter, this decrease in signal strength results directly from conservation of energy.

3. Every thought in your mind is accompanied by the motion of electrons in your nervous system. The energy to set these electrons into motion must come from somewhere. If a thought enters your mind from some outside source—whether it is through an auditory nerve or through a hypothetical telepathic transmission—that signal must be accompanied by enough energy to set into motion the necessary electrons

within your brain. For this reason, we should expect telepathic signals (if they exist) to diminish in strength as the sender becomes more remote from the receiver. But this is a principle methodically ignored by parapsychology researchers.

James Prescott Joule, during the 1840s, was the first person to do systematic experiments on conservation of energy. From that time until the middle of the twentieth century, experiments had to be performed on every kind of imaginable device or reaction to verify that the law held for every possible contingency. The number of experiments conceivable was unlimited, for in those days dozens of different kinds of energy were postulated: electric, magnetic, hydraulic, pneumatic, thermal, chemical, etc.—and it was not known where the list ended. Joule had to spin a paddlewheel in a bucket of water to measure the conversion of mechanical energy to heat. He had to send an electric current through a wire to measure the conversion of electrical energy to heat. Compressing a container of air measured the conversion of mechanical energy into pneumatic energy and then into heat. In addition to the infinite number of possible experiments, the precision available to Joule was not good. It would be surprising if he could attain better than a ten percent margin of error on his measurements of thermal and electrical energy.

The principle of inductive inference implied that to prove a general law such as conservation of energy, you had to make measurements on every possible kind of machine using every possible kind of energy. Even if you proved the law by a thousand experiments, somebody could always come along and ask the fatal question: how do you know that the next experiment might not uncover some new form of energy, or some new process that would allow energy to be created out of nothing? It was impossible to be sure that the law was absolutely true all the time. Nevertheless, making a leap of faith from the measurements available, supported by the fact that no perpetual machine had ever worked (except by fraud), most physicists did believe that conservation of energy was a universal law.

Modern physics sees the matter differently. We start with the observation that everything is made of elementary particles, and that there are only a few different kinds of particles in existence. Furthermore, everything these particles do is controlled by only a small number of forces or interactions. For example, the motion of the earth is determined by the gravitational attraction between the earth and the sun and all the other astronomical objects in the sky. The actions of two electric charges moving past each other are determined by the electromagnetic interaction between them.

These examples illustrate the basic concept underlying all of modern physics: *Everything that happens in nature can be analyzed in terms of interactions between pairs of particles.* A star, by the way, can be treated as a particle when dealing with problems of stellar and planetary motion.

For much of this century we have counted four fundamental forces as the source of all actions. These four forces, or interactions, are the basis of the standard model of particle physics:

1. The gravitational interaction
2. The electromagnetic interaction
3. The weak nuclear interaction
4. The strong nuclear interaction.

More recently we have recognized that the electromagnetic and weak nuclear interactions are two components of one force called the electroweak interaction (just as the electric and magnetic forces are two components of the electromagnetic force). This reduces the number of basic forces to three. Some physicists are optimistically trying to combine the three into a single interaction—a theory that would be a "theory of everything." Their efforts are not yet complete. For our purposes it will be convenient to retain all four as the fundamental interactions.

The point of the matter is this: when scientists test the law of conservation of energy at the present time, they do it with experiments involving interactions between the fundamental particles: i.e., between electrons, protons, photons, quarks, etc. If we find that there is no energy created or destroyed in any reaction between particles, we can be sure that energy is neither created nor destroyed in any kind of device made of these particles. We know this because the operation of the device ultimately reduces to the interactions between all its particles.

One crucial difference between a classical experiment and a modern experiment is this: each classical experiment resulted in only one measurement. A modern experiment, on the other hand, may involve an enormous number of measurements. This is because one experiment performed with fundamental particles may require the reaction of millions and millions of particles in a short period of time. Each individual reaction amounts to a separate measurement. And if a billion reactions give the same result, what is the chance that the next reaction is going to give a different result?

So particle physics offers a way to do a vast number of experiments in a reasonable period of time. Taking advantage of this, we find that whenever we bombard an atomic nucleus with neutrons, protons, electrons, or photons in order to induce a particular reaction, the amount of energy going into the reaction is very nearly the same as the amount of energy coming out of the reaction. While there are slight deviations from one reaction to the next, the limits of error are very small.

Furthermore, since there are only four different kinds of energy, and relatively few different kinds of particles reacting with each other, the law of conservation of energy can be verified by a finite and manageable number of experiments. (We will see in the next section that the number of different kinds of particles that can exist has recently been limited to a maximum of twelve.)

The problem of inductive inference posed by the philosophers is now seen to have a new answer. It is not necessary to do an infinite number of experiments to prove a generalization. A finite number of experiments will do.

It is not commonly realized how precise the experiments verifying the law of conservation of energy can be. As I have described elsewhere,[2] such experiments can have a margin of error that is almost unimaginable to those accustomed to handling a meter stick or even a digital voltmeter. In certain specialized nuclear physics experiments using a technique called the Mössbauer effect, the probable error is only one part out of 10^{15}—a one followed by fifteen zeros. That is, when we look for the change in energy taking place during a certain reaction, the Mössbauer effect is able to detect a change of energy amounting to one unit out of one thousand million million units. And within these limits no change in energy is found.

What does such fine precision mean? To give a crude analogy, suppose I am typing a manuscript at a rate of 100 words per minute. This is a fast clip, but let's assume I am a good typist. Suppose, in addition, that I am such an accurate typist that I make only one error in 10^{15} words. If I typed nonstop, it would take 20 million years for me to make one error! That is the kind of precision we can expect in measurements of energy using the Mössbauer effect.

More recently, measurements using a superconducting device called the Josephson Junction have improved the precision of energy measurements by a factor of about 3000.[3] These measurements, among the most precise in all of physics, show that energy is conserved by the electromagnetic interaction to within an error of 3 parts out of 10^{19}. If my typing accuracy improved to that extent, I would expect to make one error in about 60 billion years.

The purpose of all this discussion is to show that conservation of energy has been verified to a precision that is far greater than necessary for any practical purpose. To philosophers, of course, no amount of precision is enough. However, these physical measurements bring us exceedingly close to the "absolute knowledge" they love. Furthermore, since Mössbauer effect measurements involve fundamental particles interacting through the electromagnetic and strong nuclear reactions, our conclusion is that conservation of energy holds true (within the above limits of error) for every kind of macroscopic or man-made device that can be imagined. The reason for this conclusion is that the electromagnetic interaction is responsible for all actions taking place on a level that we personally experience: all chemical reactions, all thermal, electrical, and mechanical machinery, all solid state physics, all electronics, all biology, are determined by electrons and protons acting under the influence of the electromagnetic force. Consequently energy must be conserved during every action taken by any machine or living organism.

The strong nuclear force, on the other hand, governs the binding of particles within the atomic nucleus. It is responsible for the generation of nuclear energy and plays a large role in radioactivity. As we have seen, the strong nuclear force obeys conservation of energy to the same degree as the electromagnetic interaction.

The weak nuclear interaction also plays a role in some forms of radioactivity, and while energy conservation experiments using the weak interaction are not quite as precise as the Mössbauer effect measurements, their limits of error are good enough. For over fifty years no violations of conservation of energy have been detected in the operation of either of the nuclear forces.

The same can be said for the gravitational interaction. The fact that the planets have continued in their natural orbits for billions of years indicates that none of their energy ever gets lost (except through normal friction), and no additional energy makes itself known.

During the past century a number of observed events have appeared to violate conservation of energy and so have aroused concern. Most prominent among them was the case of the mysterious energy radiated from the element uranium. For many years after its discovery by Becquerel in 1896, it appeared that radioactivity had to imply a breakdown of conservation of energy, because an apparently inert mass of uranium put into a calorimeter proceeded to raise the temperature of that calorimeter. Furthermore, the emission of energy continued apparently without diminution over a long period of time. However, once the phenomenon was sufficiently understood, it was clear that conservation of energy was safe. The energy coming out of the uranium was simply energy that had been stored in it early in the life of the solar system. It did diminish with time if you waited long enough, or made sufficiently precise measurements.

Experiences such as this have strengthened our confidence that conservation of energy is a general law of nature verified to an extremely high degree of precision, and applicable without exception to everything happening in the universe.

The Standard Model of Matter

Some of you may object to the strong conclusion of the last section. How do we know, you might ask, that the four fundamental forces are the only forces in existence? How do we know that the known particles are the *only* particles? If there exist new and strange particles outside our ken, particles which do not obey the laws that we currently know, then it is no longer certain that the predictions of physics are always correct. Because of this possibility, we see that our ability to claim well-founded knowledge depends on the completeness of the models we make of the universe. Making a mathematical model of what is "out there," verifying this model accurately, and then showing that the model enables us to predict what is going to happen in a given set of circumstances lead us to believe that we know some things for a certainty. To show that we really know some things now that we did not know before, let us take a brief detour through the history of particle models, culminating in what we now call *the standard model of particle physics.*

The Copernican revolution that set the sun (rather than the earth) at the center of the solar system ushered in the age of modern science. However, the Copernican model is more geometry than physics. The sun works better at the center of the solar system than at the circumference because the diagrams describing the motion of the planets are neater. One does not have to work as hard to account for the weird back-and-forth motion of Mars. But the Copernican idea does not really add to fundamental understanding. It is a descriptive model only and does little to satisfy our curiosity about *why* things are the way they are.

Embarrassing questions followed in the wake of Copernicus. Keeping in mind Aristotle's edict that heavenly bodies need prime movers to keep them going, the first logical question was: why does the earth continue to move around the sun? And: why do the planets farther away from the sun travel more slowly than the planets closer to the sun? Why does a model consisting of simple circular orbits fail to agree with the observations of planetary motion?

After Copernicus, another half century elapsed before Johannes Kepler, examining the observations of Tycho Brahe, decided that the last question could be answered by assuming that the orbits of the planets were actually ellipses, rather than circles. But that answer simply led to new questions. Why is the earth's orbit an ellipse rather than a circle? Indeed, why does it travel in a curved path at all? It was becoming clear that science needed a way to answer questions that asked *why* things happened, instead of simply finding better ways to describe what was happening. But another century had to pass before Isaac Newton learned how to answer the "why" questions, putting his reasoning into the book that really began modern physics—the *Philosophiae Naturalis Principia Mathematica* (1687).

Newton's monumental creation was essentially the first theory of particle physics. The idea consisted of two parts:

1. Newton thought that a planet or satellite ought to travel in a straight line unless there was some kind of force compelling it to move in a curved path. (Newton's first law of motion.) Since the planets do not travel in straight lines, there must be some sort of force pulling each planet into a curved path.[4]

2. The famous falling apple hinted to Newton that the force acting on the moon and the planets was the same gravitational force that acted on objects at the surface of the earth. This had not been suspected

before. He then put into use his second law of motion—the law that tells how to calculate the motion of an object with known forces acting on it. Knowing how fast the moon went around in its orbit, Newton was able to compare the strength of the force acting on the moon with the force acting on his apple, and so deduced that the gravitational force has to follow an inverse-square law. No explanation was given as to how the force came about and how it was transmitted through space. It was simply a requirement of the model, verified by observation, perhaps to be explained on a deeper level in the future.

The whole theory of planetary motion held together because Newton was able to prove mathematically that, when using the inverse square law of gravitation to calculate the force between two spherical bodies, it was all right to treat the problem as though all the mass of each body was concentrated at its center. Thus it was legitimate to treat the sun and the planets as though they were point particles. It was this theorem that guaranteed the success of the Newtonian model, for without it the theory would have been too complex for application in the precomputer age.

Thus arose the great new paradigm of physics: the concept that the motion of all the planetary bodies could be explained as resulting from the gravitational forces acting between *pairs* of objects. It didn't matter how big these objects were: they could be as large as the sun or a planet, or as small as an apple. The same force worked on all of them. The gravitational force was the shaper of all trajectories, determining the paths of stars, planets, cannon balls or baseballs.

The ultimate success of Newton's model depended on observations of planetary orbits. The more accurate the measurements, the more confident were scientists that the model was correct. The history of science since Galileo and Newton has been a history of improvements in instrumentation and in the accuracy of physical measurements. Modern measurements have become so precise that it becomes necessary to make corrections for effects that Newton never thought of. In computing the orbit of a man-made satellite around the earth we must take into account such infinitesimal perturbations as the gravitational pull from other planets, the slight friction of the tenuous atmosphere out in space, the fact that the satellite might have picked up a few electrical charges which would affect its orbit as it moves through the earth's magnetic field, and the fact that the earth is shaped like a pumpkin rather than a perfect sphere.[5]

When all these corrections are made, it is found that the planetary orbits agree with Newton's theory to better than one part out of 10 million. This is good enough to give us great confidence in the theory's validity. However, pride goeth before a fall. When the precision of the measurements gets better than one part out of 10 million, small discrepancies begin to show up. The orbit of the planet Mercury, for example, differs ever–so–slightly from the predictions of Newton's theory. These tiny discrepancies were first discovered by the French astronomer Urbain Jean Joseph Le Verrier in 1859, but remained unexplained for another 60 years.

To find an explanation required a complete reworking of the theory of gravitation and the laws of motion. When this was accomplished by Albert Einstein with his general theory of relativity in November of 1915, Le Verrier's observations gained an explanation that required no further assumptions.

Einstein's theory of gravitation has been verified with a precision allowing us to claim a fit between observation and theory with a possible error of about one part in a billion. It becomes difficult to make a better fit because we are uncertain of the precise shape of the sun. We know the sun to be in the form of an oblate spheroid, but the shape we see on the outside may not be quite the same as the shape on the inside, where most of the mass is located. After all, the sun is not a solid object, and different parts of it rotate at different speeds. As a result, its precise influence on the orbits of the planets is difficult to determine.

I emphasize the high precision of our measurements of planetary motion because, as mentioned previously, our knowledge is only as good as the quality of the measurements verifying that knowledge. Our present model of the solar system has been verified to an exceedingly high degree of precision. Therefore, if the model is ever modified in the future, these changes can make no more than small corrections to large-scale observable phenomena, and will not diminish our ability to use the model in calculating the orbits of planets, comets, stars, galaxies, and space vehicles. Our ability to send a vehicle past Uranus and Neptune is proof that we do know the solar system quite well, if not with absolute perfection.

The planetary model is the template for the atomic model, in which each atom of matter is akin to a tiny solar system, with electrons surrounding a massive nucleus containing neutrons and protons. The atoms, in turn, join to form molecules and crystals, which in turn form solids, liquids, and gases. Thus, they make up all of matter. The gravitational force, so important in determining the dynamics of the solar system, is utterly insignificant within the realm of the atom. This is because the electrons and protons, being electrically charged, interact by means of the electromagnetic force, and the electrical attraction between an electron and a proton is about 10^{39} times stronger than the gravitational attraction between them.

The rules of the electromagnetic interaction are very well known. Some of the most precise measurements in physical science have validated this knowledge.[6] The fact that the electrostatic attraction or repulsion varies with distance according to the inverse-square law (Coulomb's law) has been experimentally verified to a precision of about two parts out of 10^{16}. The fact that the electron has exactly the same amount of electric charge as the proton (although of opposite sign) has been verified to an ever greater precision—about four parts out of 10^{20}. The latter figure is an example of the extremely high sensitivity of experiments designed to show that two electric charges are equal. There are fewer uncertainties in showing that two charges are equal than in trying to measure their magnitude.

During the first half of the 20th century physics struggled through an explosion of new discoveries concerning the particles that make up the atom. These discoveries were of three types:

1. There was a proliferation of new particles. The electron, proton, neutron, and photon seemed mundane compared with the positron, the neutrino, and the dozens of types of mesons that appeared in particle accelerators when the available energy became great enough. The mesons were short-lived particles first found in cosmic rays with masses between those of the electron and proton. However, with the advent of very high energy accelerators, a large variety of particles were found with masses greater than that of the proton, and the Greek alphabet became exhausted in naming them. These heavy particles became known as *baryons*.

2. New types of forces made their appearance. In addition to the classical gravitational and electromagnetic forces, two new kinds of forces

were found to operate within the atomic nucleus: the weak nuclear and the strong nuclear forces. More important, every kind of reaction that took place between the fundamental particles appeared to be explainable with the help of those four forces alone.

3. Newton's laws of motion were found to be useless when trying to understand how elementary particles interact with each other. Nor do they work when we try to calculate the shapes of individual atoms or the behavior of the atoms within bulk matter. In dealing with matter at or below the molecular level we must use the mathematical apparatus of quantum mechanics, in which matter exhibits properties more like those of waves than the classical properties of particles.

But quantum theory does more than supply recipes for computing the form and behavior of matter. The branch of quantum theory known as quantum electrodynamics (QED) gives an explanation for the electromagnetic force in terms of deeper concepts. For the first time in history physicists are able to answer questions about the underlying causes behind the existence of the four fundamental forces. It is no longer necessary to accept action at a distance as an unfathomable mystery.

Within the theory of QED the electromagnetic interaction is found to result from the exchange of photons between pairs of charged particles. Electrostatic attractions and repulsions, magnetic effects, and radiation can all be derived from the equations describing photon exchange between charged particles. In addition, this photon exchange introduces subtle changes in the wavelengths of light emitted by the hydrogen atom (the Lamb shift). Measuring these changes provides an experimental verification of the theory.[7]

Actually, every experiment in physics involving the dynamics of charged particles provides precise verification of electromagnetic theory. As a result of work done over the last century, the behavior of electromagnetic fields has been verified down to a distance of 10^{-16} cm, much less than the diameter of an atomic nucleus. The predictions of this theory have been experimentally verified to within one part in a million. Therefore the model we use when we visualize the electromagnetic force as resulting from the exchange of photons between charged particles is not "just a theory." Quantum electrodynamics is the most successful, complete, and precisely verified dynamical theory in the history of science.

Following the success of the QED model, scientists have generalized the theory to explain each of the four fundamental forces as arising from the exchange of a specific type of particle. Since there are four forces, there must be four different kinds of exchange particles, or intermediators:

1. The photon is the particle that mediates the electromagnetic force between electric charges.

2. The gluon is the particle that mediates the strong nuclear force that holds quarks together to make neutrons and protons, and which holds the neutrons and protons together within atomic nuclei. (The gluon is named after the nuclear "glue" that binds nuclear particles together.)

3. The vector boson is the name given to the particle responsible for the weak nuclear force. To be precise, there are three kinds of vector bosons, corresponding to three kinds of weak forces: these are the W^+, the W^-, and the Z^0. The mathematical theory describing these forces shows that they are closely related to the electromagnetic force. Therefore, instead of considering the electromagnetic and weak forces to be two separate kinds of forces, we know that they are actually two aspects of a single force: the electroweak interaction.

4. The gravitational force is the interaction between masses. It is a long range force, follows the inverse-square law, and it is extremely weak relative to the other forces. In quantum theory the gravitational force arises through an exchange of hypothetical particles called gravitons. These gravitons have not yet been observed and the theory is incomplete, but as far as observable properties is concerned, gravitation is well understood. (The general theory of relativity still stands as an alternative way of understanding gravitation, but this theory breaks down under extreme conditions such as those in black holes and at the beginning of the "big bang." A union of quantum theory with general relativity is expected to remove these difficulties.)

The most important aspect of the new particle physics is the way the multitude of particles discovered during this century has been reduced to a more manageable number of basic entities. These particles (which we tentatively label as "fundamental") have been organized into three groups, each with four different kinds of particles. This advance was made possible by the invention of quarks in the second half of the century. I use the word "invention" rather than "discovery" because the quarks are truly intellectual inventions. Individual quarks cannot be seen or detected in any manner. However, quark theory makes specific predictions about the behavior of the observable particles, and these predictions can be verified. For example, we can show how the neutron, proton, and all the mesons are composed of combinations of several kinds of quarks, and we can calculate the masses of those particles from the theory. Because of this we have discarded the old positivistic rule that "we don't talk about anything unless we can measure or detect it." The new rule is: "We can talk about unobservable things as long as they produce predictable and measurable effects."

Since neutrons, protons, and mesons are composed of quarks, they are no longer considered to be elementary particles. The known fundamental particles that remain are divided into two types:

a. Quarks (of which there are six kinds) are heavy particles which do not exist separately, but which are always found in groups of twos or threes. Pairs of quarks make up mesons, while triplets form baryons, the class of numerous heavy particles whose most recognizable members are the neutron and proton.

b. Leptons are light particles, such as the electron, the muon, the tau particle, and three kinds of neutrinos.

The first thing we notice about this grouping is that there are six kinds of quarks and six kinds of leptons. It has been found useful to rearrange the particles into three classes (or generations, or families) each with two quarks and two leptons:

1. Quarks: u (up) and d (down)
   Leptons: electron and electron-neutrino

2. Quarks: c (charm) and s (strange)
   Leptons: muon and muon-neutrino

3. Quarks: t (top) and b (beauty)
   Leptons: tau and tau-neutrino

For each of the particles above there is a corresponding anti-particle. The positron, for example, is an anti-electron. However, in our universe anti-particles are found only under special conditions. They are not part of ordinary matter.

In fact, all the particles found in ordinary matter are made up of the occupants of family 1 in the above listing. The particles in families 2 and 3 are exactly like their counterparts in class 1 except that they have greater masses. (For example, a muon is exactly like an electron, except that its mass is about 200 times greater.) The class 2 and 3 particles come into existence only at very high energies.

The description above is the barest outline of the "standard model of particle physics." It is the model accepted by scientists as the basis for their work in all areas of natural science. The literature on the standard model is extensive and rapidly expanding, and it is not my intention to repeat it here. My purpose, rather, is to ask whether our standard model is sufficient to explain everything we can observe. Or, to turn the question around: do we need more than quantum theory, the presently known particles, and the four fundamental interactions to explain the solar system, the operation of a computer, the structure of both living and non-living matter, the mechanism of evolution, and the origin of human consciousness?

This question can be broken down into two parts:

1. Do we know all the particles in existence?

2. Do we know all the forces which account for the activities of these particles?

As far as the first question is concerned, recent developments in particle physics allow us, for the first time in history, to put a limit on the kinds of particles that can exist. Experiments performed with high-energy accelerators, beginning in August 1989, have determined that the number of different particles that can exist are no more than the three classes listed previously.[8] The reasons are not simply an inability to find more particles. The new experimental results actually put a limit on the number of different particles that exist.

These experiments put a new complexion on the limits to the standard model. The key question is no longer: "Are there new particles lurking somewhere in the universe that have not yet been discovered?" Since elementary particles must come in families, the question becomes: "Are there new classes of particles waiting to be discovered?"

What we can say now with a high degree of assurance is that the three families of particles we listed earlier are all-inclusive. The initial experiments of 1989 allow us to claim that only three classes of particles exist, and this claim is made with a confidence level of 98 percent. Further experiments will undoubtedly improve the statistics and increase our confidence. One proviso must be made, however: there could theoretically be a fourth family, but its neutrino would have to possess a mass greater than 30 GeV. Otherwise it would already have been detected. (The proton mass is about 0.94 GeV.) A neutrino more massive than a proton is considered very unlikely, since each of the first three neutrinos has zero mass (or very close to zero). At any rate, particles belonging to such a massive family could have no relevance to actions taking place in matter under ordinary conditions.

The question about the possibility of new forces is related to the question about new particles. In particle physics, a force is an interaction between a pair of particles. The four known interactions account for all of the interactions observed among the particles of the three families. It cannot be denied that forces might exist that are so weak that they have until now avoided notice. But what of new and unsuspected strong forces? We will come back to this question when we analyze Myth 4.

In the meantime, the standard model of particle physics represents, to a large degree, knowledge that we "know for sure." This is not to say that the standard model is complete. There are still an enormous number of questions to be answered. We would like to know, for example, why the electron and proton have exactly the same amount of electric charge. We would like to know if the electron has a structure —if it is composed of simpler particles. We would like to know how to calculate the masses of the various particles from basic principles. Overshadowing these detailed questions are the philosophical mysteries of quantum theory.

The issue of perfect knowledge incorporates questions concerning how much we know, how much we don't know, and knowing how well we know the things we know. Paradoxically, understanding uncertainty improves the certainty of knowledge.

The Danger of Modesty

Dogmatists of all types occupy the trenches of the counter-scientific culture. When scientists develop an air of objectivity in order to avoid the appearance of arrogance, they run the risk of playing into the hands of their enemies. By saying that "all knowledge is uncertain," the biologist hands ammunition to the creationist, who argues with perfect confidence that his creationism theory is just as good as the theory of evolution. The fact that evolution is far more satisfactory as a theory than creationism gets lost in the heat of the argument. By saying "we don't know anything for sure," the scientist leaves himself without a defense against the theories of the UFO, ESP, and astrology enthusiasts.

The word "theory" has itself become a weapon in the wars of scientific heterodoxy. In popular usage, a theory is little more than an hypothesis, or a guess. But to a scientist a theory is an elaborate structure of observation, hypothesis, general principles, deductions, mathematical analysis, and experimental verification. To a physicist, the term "kinetic theory" represents a system for calculating the properties of gases starting with the equations of motion of the gas molecules. There is no guesswork involved. To a physicist, the term "quantum theory" represents a set of equations from which the properties and behavior of all particles, atoms, molecules, and radiation can be calculated. This theory is the most thoroughly verified system of knowledge in history. At this state of the game there is no doubt that the theory, though incomplete, does work. Chemists can now use quantum theory to calculate the structure of reasonably complex molecules as well as the results of chemical reactions, starting from basic principles and working up from the bottom.

Even in newly developing fields such as high temperature superconductivity, where we do not as yet know how to explain the observable effects, there is no doubt that the basic principles of quantum theory apply to the problem. The mysteries encountered simply revolve around the problem of setting up a new model to represent a new and complex type of system.

Similarly, the biologist views "evolutionary theory" as a set of principles based on a great deal of biological and geological evidence and has no doubt that eventually all biologists will agree on some version of evolution that will explain all the observations. Evolution is not just a hypothesis. Though there may still be controversies concerning the specific mechanisms of evolution, this is to be expected in a field that is still developing.

Because of the many opportunities for misunderstanding in communicating with the public, it is often best to dispense with the word "theory" entirely. Too often relativity has been dubbed "just a theory" by its opponents. For this reason physicists prefer to use the term "principle of relativity" rather than "theory of relativity."

The professional scientist is not satisfied only knowing the content of a theory. He must also know how the theory originated, what experimental evidence verified it, and how complete that evidence is. He must also know what part of the theory is hypothesis, how much of it is opinion, and how much of it is truly validated knowledge. Most important, he must be aware of how any given piece of knowledge entered his mind. Only with this kind of knowledge can the scientist know what the limits of his certainty are.

It is hard enough for the professional scientist to be totally aware of all the details related to his work. What is the general public to do? Science writers, eager to emphasize the strange and exotic, titillate the reader with speculations of portentious discoveries just around the corner until the reader is unable to separate fantasy from reality.

Quantum strangeness has spawned countless books promising telepathy, faster-than-light communication, time travel, and other wonders emerging from the realities of the New Age. And yet, the scientists most personally involved in research on this topic—people such as the theorist J. S. Bell, or the experimentalist Alain Aspect—are adamant in their rejection of faster-than-light communication.

It becomes apparent to the science–watcher that attitudes toward a particular theory are likely to be based on unconscious and irrational biases, rather than on empirical evidence or reasoned argument. Those prone to pragmatism want a theory to make specific predictions; the predictions can then be checked to ensure that the events predicted actually happen. Those of a mystical bent, on the other hand, are less interested in being able to reason rigorously from cause A to result B.

The parapsychology enthusiast, for example, believes that he can demonstrate examples of ESP or precognition. But he has no system of causality to explain how these events occur. He has no carrier for the messages that travel from one mind to another, no way to explain how information travels from the future to the present. The New Age follower, for instance, has a theory that reincarnation can explain why people behave the way they do. But with this kind of theory you can explain anything you want. There is no way of using the theory to make a specific prediction of observable events.

Verifiable predictions are the mark of an empirical scientific theory. Classical mechanics enables us to predict where the earth will be a year from now. Quantum mechanics enables us to predict the wavelengths of light emitted from a given atom. This predictability is what makes science an empirical system of knowledge.

Conversely, if a theory is unable to make predictions of observable effects capable of verification, then it makes no difference whether or not you believe it—except insofar as the belief itself has an effect on your feelings and actions. The main function of nonempirical, mystical theories is to make people feel good.

However, since non-empirical theories often cause physical and mental damage, as when faith healers promise cures that they cannot deliver, or when creationists try to force bad science and bad thinking into our schools and text-books, it is important for scientists to remember how much they really do know, and to stop apologizing for it.

NOTES

1. J. Bronowski, *The Ascent of Man* (Boston: Little Brown & Co., 1973), chap. 11.

2. M. A. Rothman, *Discovering the Natural Laws* (New York: Doubleday & Co., 1972, and New York: Dover Publications, 1989), chap. 6.

3. D. G. McDonald, "Superconductivity and the Quantization of Energy," *Science* (12 Jan. 1990): p. 177.

4. René Descartes actually preceded Newton in the statement of the first law of motion. In *Principia Philosophiae,* published in 1644, he says, "Every body tends to continue its motion in a straight line, not a curved line, and all curvilinear motion is motion under some constraint." See B. Russell, *A History of Western Philosophy* (New York: Simon & Schuster, 1945), pp. 561–562.

5. M. A. Rothman, *Discovering the Natural Laws,* chap. 4.

6. Ibid., chap. 8.

7. S. J. Brodsky and T. Konoshita, "Quantum Electrodynamics," in *The Encyclopedia of Physics,* 3rd ed., R. M. Besancon, ed. (New York: Van Nostrand Reinhold, 1985), p. 983.

8. J. Kinoshita, "The Lost Generation," *Scientific American* (December 1989): p. 22; SLAC Staff, "Measurements of Z-Boson Resonance Parameters in e^+e^- Annihilation," *Phys. Rev. Letters* (13 November 1989): p. 2173; CERN Staff, "Determination of the Number of Light Neutrino Species," *Physics Letters* (16 November 1989): p. 519.

MYTH 3

"Nothing is impossible."

The Meaning of Impossible

> **possible** 1. that can be; capable of existing; 2. that can be in the future; that may or may not happen; 3. a) that can be done, known, acquired, selected, used, etc., depending on circumstances (a possible candidate); b) that can happen or be; potential; 4. that may be done; permissible; 5. that may be a fact or the truth; 6. [colloq.] that can be put up with; tolerable.

> **impossible** 1. not capable of being, being done, or happening; 2. not capable of being done easily or conveniently; 3. not capable of being endured, used, agreed to, etc., because disagreeable or unsuitable (an impossible task, an impossible request).

The statement that "nothing is impossible" is equivalent to saying that "everything is possible." This is an aphorism heard often, especially in the form "Anything is possible if you just try hard enough." It is a beloved gem of teachers, coaches, cheerleaders, and other devotees of positive thinking. Undoubtedly the saying helps encourage underachievers to test their limits. You never know how far you can go until you try your hardest. But reality requires us to recognize that everything of a practical nature does have limits.

Counterexamples to the myth that "nothing is impossible" abound:

a. I know that I will never play the Liszt piano sonata the way Vladimir Horowitz did. (I'll never play it the way anybody does, for that matter.)

b. The production of energy in the world is not going to increase at a rate of 2 percent per year indefinitely.

c. Nobody is ever going to put a kilogram of water (in liquid form) into a 100 cubic centimeter beaker.

d. Nobody is ever going to express *pi* (the circumference of a circle divided by its diameter) by a number that has a finite number of decimal places.

e. Nobody is going to build a machine that puts out more energy than it put into it (by heat, electrical power, burning fuel, etc.)—even though at least two people are currently claiming they have done it.

f. Nobody is ever going to build a vehicle that floats motionless thousands of feet in the air without mechanical or otherwise visible means of support. For this reason nobody is going to prove beyond reasonable doubt that they have seen a UFO suspended in the sky with only an antigravity device to hold it up.

g. Nobody is going to build a spaceship that can travel to another star in less than a year's time.

h. Nobody is ever going to send telepathic messages to the moon.

Each of these examples demonstrates the mythology of the aphorism "nothing is impossible." The word "nothing" allows for no exceptions. Yet I have, with little difficulty, conjured up several counterexamples proving the saying to be false.

To understand why the actions listed above are impossible, we must talk about the many different meanings of the world "impossible." We will see that some of our impossibilities are trivial, while others are quite serious, from a philosophical, scientific, or practical point of view.

a. Conditional impossibilities are actions that are impossible within a specific set of circumstances. While it is impossible for anybody to fly to the moon without some kind of mechanical propulsion, this is not a generalized impossibility, for we know that people can go to the moon with the aid of the right machinery.

b. Resource impossibilities are things that cannot be done because the necessary resources are not available. It is impossible for me to buy the General Electric Corporation because I don't have enough money or credit to do so. For somebody else it might not be impossible.

c. Material impossibilities are based on the properties of substances. Because of the way water is formed, it is impossible to put a kilogram of liquid water into a 100 cubic centimeter beaker. Normally a kilogram of water occupies 1000 cm^3. If you tried to squeeze it into a much smaller space, you would end up with a solid.

d. Mathematical impossibilities are based on the properties of numbers and other mathematical entities. These are generally based on rigorous definitions. Because of the definition of real numbers, it is impossible to add one apple to one apple and get four apples. If you try changing the definitions, you end up with a different kind of mathematics, one which may not apply to real objects.

e. Fundamental physical impossibilities are imposed by the structure of the universe. These are actions that are forbidden by nature. Things cannot do what they are not able to do, no matter how hard humans try to make them do so. Humans have absolutely no control over nature when they try to defy the way the universe works. This rule is especially appreciated by experimental physicists who struggle with recalcitrant equipment that insists on having its own way. The focus of this book will be on these physical impossibilities.

Conditional Impossibilities

It's perfectly true that it is impossible for me to play the Liszt piano sonata because my nervous system doesn't have the speed and memory required for this gargantuan task, and because my muscles are not trained for it. It may be that no amount of training will ever put me in shape for the job. Other people, however, can and do play this piece of music although very few can do it the way Vladimir Horowitz did. Therefore, this is a conditional impossibility—not a general impossibility.

Often when coaches and educators attempt to encourage their charges by saying, "Anything is possible if you try hard enough," what they really mean is: "You don't know how much you can accomplish if you don't give it your best effort." If, on the other hand, they really mean that anything is possible, they are deliberately ignoring individual differences. A high school teacher who thinks that everybody can learn calculus with equal ease is making a grave error. Some students will soak up calculus like a sponge, while others will require years of study to comprehend the abstractions involved. Whenever a study requires more than a certain amount of effort, fatigue and boredom begin to set in. A class in which everybody is forced to move at the same rate either is going to move very slowly, or is going to end up with a lot of dropouts.

A reading of musical and mathematical biography has convinced me that some people have nervous systems that can do things that are impossible for others. Many mathematicians (some of whom are self-educated) can perform mental calculations that you or I cannot do even on paper. Charles Proteus Steinmetz, the wizard of General Electric who contributed much to the development of electrical engineering in the early part of the century, had a habit of doing his theoretical research while he was in a rowboat in the middle of a lake. If he forgot his table of logarithms (a necessity in the days before laptop computers, or even pocket calculators) it didn't matter to him. He would calculate these long numbers in his head.

Even more astonishing was the career of the Indian mathematician Srinivasa Ramanujan, who was born in 1887, was raised in a poor family, had a minimal education, and during the course of this education read only a few elementary books on mathematics. From that starting point he worked out much of nineteenth-century mathematics entirely on his own, proving all the theorems and solving complex problems by his own methods. Could he have progressed further and faster had he gone to Oxford? After making contact with G. H. Hardy, Ramanujan was brought to England by that eminent mathematician, with whom he collaborated until he died at the age of 33.

Often the greatest musicians have shown their extraordinary abilities by the age of four or five. Artur Rubinstein, at the age of three, would listen to his older sisters practice their piano lessons, and afterwards sit and play the same pieces from memory.

The exhortation "anything is possible if you try hard enough" may save a number of late bloomers, but the other side of the coin is the number of disappointed strivers who discover that they do have limitations, and no matter how hard they try, they are not going to make the Olympics, or play like Artur Rubinstein. To put it another way, it is not possible for everybody to be the world champion at the same time.

Other kinds of conditional impossibilities are related to the inadequacies of current knowledge, or the availability of the right tools. Fifty years ago it was impossible for anybody to compute the number *pi* to a million decimal places because the computer power was not available, and nobody could have lived long enough to do it by hand. Now the job is readily done. Impossibilities of this nature are local or temporary.

At present it is impossible to build a computer that thinks like a human being. This is simply because we do not know how to do it. Is this task truly impossible? As far as the future is concerned, it is too soon to say whether it is possible or impossible. We are now at the point where computers can beat most humans in chess. However, this kind of dumb computation (or competition) has little relation to the way humans actually think. Furthermore, the numerical and logical abilities of present-day computers do not begin to approach the more subtle and complex manifestations of consciousness, sensations, and emotions.

Nevertheless, it is premature to rule out the possibility of building a computer that does indeed possess consciousness. If the human nervous system can create consciousness out of a network containing a finite number of cells, then what is there to prevent a complex network of transistors or equivalent devices from doing the same thing? If it turns out that it is indeed impossible to create a computer that is aware of its own existence, this failure would imply that something is required in the nervous system over and above the existence of normal atoms and molecules responding to natural forces. Would this extra ingredient be something mystical, psychic, or supernatural; or would it be some new and unknown process within the domain of science? Because of this question, the possibility or impossibility of conscious computers is one of the central questions in modern science.

It may well be impossible to build a thinking computer simply because of the complexity of the job. Do humans have the capacity to assemble billions of transistors into the necessary arrangement within a finite period of time? If the answer is negative, *not* being able to build a computer with human capabilities would tell us little about the problem of consciousness. Being able to build one would tell us a great deal.

Resource Impossibilities

Some actions are impossible because the necessary resources do not exist. It is impossible for the population to grow at a steady percentage rate for an indefinite period of time because sooner or later there will be more people than there are square meters on the surface of the earth. Some people might say this is going to take forever, so why worry? Statements such as this are a sign of innumeracy, because we can easily calculate how long it is going to take for the earth to become saturated with humans by putting the right numbers into the right equations. We know that the surface of the earth (including oceans) has an area of about 500 million million square meters. The present population of the earth is about 5,000 million. If the population grows at a rate of one percent per year (which offhand seems like a very modest figure), then the number of people in the world would double every seventy years. But at this rate there will be one person for every square meter of earth surface in about 1,200 years, which is not very long from a historical perspective.

Of course, long before the earth gets saturated with people, those unfortunates who do survive will run out of places to store garbage, or farmland on which to grow food. Yet many well-meaning people persist in believing that the birth of more humans is the greatest thing in the world.

Increase of population is accompanied by increase in energy consumption and production. We tend to think how nice it would be if energy production was so easy and cheap that it could continue to increase indefinitely. However, energy production is limited by an interesting and practical consideration: regardless of how energy is generated, and regardless of the form in which it first appears, it always ends up in the form of heat. This is a consequence ignored by those who look with an optimistic eye at the glorious future. The recent hysteria over "cold fusion" was prompted by thoughts of the almost infinite amount of energy stored in the deuterium of the earth's oceans. However, if such a cheap source of energy were easily available, efforts at energy conservation would be discouraged, and growth of energy consumption would be the inevitable outcome. If energy production grew at the same rate as the population (1 percent each year, using the example given above), it would double in seventy years. And if it kept doubling and redoubling at the same rate, in about one thousand years the amount of power generated on earth would equal the power the entire planet receives in the form of light and heat from the sun. The accompanying rise in planetary temperature would make the greenhouse effect look like air conditioning. Nothing on earth can grow indefinitely.

Another type of resource problem arises in connection with finances. Many theoretically feasible projects collapse under the weight of potential cost. The literature of science fiction is filled with gargantuan construction projects, and in science fiction nobody ever asks how much anything costs. But if—let us say—an interstellar spaceship would cost more than the total world income for a decade, it is unlikely that the device would be built with the consent of the population. At present we can't even raise the money for necessary projects such as rebuilding crumbling roads or moribund inner cities, and the projected costs of a simple space station is beginning to raise eyebrows.

One possible exception to this scenario might be found if the sun was about to explode. Then the spaceship would be deemed absolutely necessary for the survival of the human race, and then perhaps the resources could be found. (If you could get the politicians to believe it, that is.)

Here is a resource impossibility of another kind: consider the problem of building a complex system consisting of thousands of computers programmed to operate hundreds of machines required to operate at the extreme limits of physical tolerance. This system has never been tested as a whole in a realistic setting. Furthermore, it must sit unused until an emergency takes place, whereupon the entire system must spring into action and work perfectly the first time.

This is what was proposed under the rubric of "Strategic Defense Command," known popularly as the Star Wars defense system. Anybody with laboratory experience in getting new equipment to work properly has an instinctive feeling for the absurdity of such an idea. Furthermore, the system was proposed at a time when it was not even known whether the core of the proposal (the x-ray laser) would work as advertised, because it had never been tried. And when it was tried, the initial experiments gave falsely optimistic results because of experimental errors. It was for reasons such as these that most working scientists opposed the idea from the beginning.

Material Impossibilities

No matter how hard you try, you cannot make a piece of matter do what its structure and properties will not allow it to do. A kilogram of water under normal conditions occupies a volume of one liter. You can increase or decrease that volume by changing the temperature or pressure, but if you try to squeeze the liquid far enough to jam a kilogram into 100 cubic centimeters, you will no longer have liquid water. You will have a solid of some kind.

Is this example too trivial? Then consider the serious question of miniaturizing computers to the smallest possible volume. A frenzy of activity aimed at this end is being carried on throughout the computer industry as scientists try to cram as many circuits as possible into a tiny package. A serious question concerns the maximum number of transistors (or equivalent items) that can be put on a single memory chip. This question is complex and has an uncertain answer. However, it is easy to state a number that gives the absolute maximum number of units that can be packed into a given volume of crystalline silicon. This number is: 6×10^{23} units per 28 grams of silicon. This means that no more than six hundred thousand million million million units can go into 12 cubic centimeters of silicon.

How do we arrive at this number? It is simple. We learn in freshman chemistry that a gram molecular weight of any element contains an Avogadro's number (6×10^{23}) of atoms. We also know that silicon has a molecular weight of 28, so a gram molecular weight of silicon is 28 grams. You cannot possibly have a transistor with less than one atom in it, so the number above is an upper limit for the number of transistors you can get in the given crystal. Actually, this number is very far from the practical maximum, since it takes many millions of atoms to make a transistor or similar device. So the number I have given is a gross overestimation of the truth. However, it illustrates how arithmetic may illuminate the absolute limits to possibility.

Understanding the properties of matter allow us to put all kinds of limits on the kind of machines that can be built. For example, it would be nice to have a solid material that can withstand a temperature of 10,000 degrees Celsius. (By comparison, the surface of the sun has a temperature of about 6000° C.) Unfortunately, at such a high temperature the atoms of any material are moving about so rapidly that their motion is too much for the attractive forces holding them together. (Putting it technically, the kinetic energy is greater than the potential energy of the binding force.) Therefore, it is impossible for any material to remain a solid at a temperature of 10,000° C—at least not at normal pressures.

Similar arguments determine the greatest strength possible for solid materials. The tensile strength of a solid is the amount of force per square centimeter needed to tear it apart when you pull on it. This strength is determined by the forces between the atoms of the material, and there is not much you can do about these forces except choose materials that have the maximum strength. A good steel has a tensile strength of about 7,000 kilograms per square centimeter. One of the strongest materials known is a single elongated crystal of graphite, with a tensile strength of roughly 200,000 kg/cm². This graphite crystal, bonded with various plastics to provide strength against bending, is often used to build light-weight aircraft and other structures. It is thirty times stronger than steel. Can we expect to gain another factor of thirty in the future? Is steady progress possible?

The answer is: probably not, if the graphite crystal is beginning to butt up against the ultimate limit of interatomic forces.

Material limitations of this type set limits to every kind of engineering proposal. Mechanical structures must be strong enough to withstand the stresses that are inevitable. Therefore, there is an irreducible mass required for these structures. No wheel can turn faster than a certain speed; if it did it would tear itself apart by centrifugal force.

Science fiction is fond of inventing materials with strengths incomparably greater than those of existing substances. Unfortunately, every imaginable material must be governed by the forces between the atoms of which it is made, and we know all the different kinds of atoms in the periodic table.

Some authors attempt to dispense with atoms altogether and have conceived a material called neutronium, made of neutrons packed very close together. Such a material would have a density of about 100 million million (10^{14}) times that of water, like the matter found in the interior of nuclei, or in the center of stars. Unfortunately, the inventors of neutronium failed to notice that neutrons by themselves are somewhat radioactive, with a half-life of 12 minutes. Inside an atomic nucleus a neutron is stabilized because of its proximity to a corresponding proton. But any nucleus that has many more neutrons than protons is found to be unstable and quickly gets rid of enough electrons to even things out. Thus the thought of making a spaceship out of an impenetrable shield of neutrons is the fantasy of an author who does not like to let a few facts get in the way of a good story.

Mathematical Impossibilities

Mathematics is a purely human creation. Mathematicians start with axioms and postulates and deduce complex structures of logic. Once a mathematician defines a real number, or a complex number, or a straight line, there is no way of arguing with the definition. One item may even have more than one definition of equal validity. There are several definitions of parallel lines, for example. Each represents a different kind of geometry. Whether or not the object defined corresponds to anything existing in nature is a separate question. That is where we move from mathematics to physics.

For example, an imaginary number is a number that has the square root of minus one in it. What an absurd notion! In ordinary arithmetic such a number is impossible. But in the arithmetic of complex numbers, the square root of minus one is a well-defined entity, with well-defined rules that tell us how to do calculations using this number. Still, one would not expect to find anything in nature that corresponds to such a strange idea. It must be nothing more than an abstract notion. However, in spite of their strangeness, imaginary numbers are really quite useful in describing physical phenomena that have wave properties: sound waves, light waves, alternating electric currents, etc. Indeed, the entire mathematics of quantum theory is based on the properties of imaginary numbers, so the term "imaginary" is a semantic deception. Imaginary numbers are just as real as "real" numbers.

Mathematicians are very fond of *rigor*. A truly rigorous proof of a theorem cannot be disputed. The proof starts with well-defined terms and proceeds with well-defined operations to an unambiguous and inarguable conclusion. A large part of the development of modern mathematics is associated with the invention of methods for improving the rigor of mathematical proof.

Over the years mathematicians have proven a number of theorems with widespread repute outside the confines of academia. Two of the most famous are the following:

1. *Squaring the circle:* it is impossible, using only a straightedge and compass, to construct a square with an area exactly equal to that of a given circle.[1] Equivalently, it is impossible to calculate the exact circumference of a circle, given the diameter. This matter, in turn, is equivalent to the well-known fact that the number *pi* is a transcendental number and cannot be written with a finite number of decimal places. Of course, *pi* can be written out with as much accuracy as might be needed for any practical purpose. (A hundred decimal places should do very nicely.) But that is not the point. Mathematicians are interested in the *theoretical* proposition of squaring the circle. Theoretical exactness is not the same as practical exactness.

2. *Trisecting the angle:* it is impossible to divide any given angle into three equal parts, using only a straightedge and a compass. The reason for this has to do with the impossibility of representing a cube-root by geometrical operations that employ only a ruler and compass.

The most fascinating aspect of all this is the exaggerated response of certain non-mathematicians to the restrictions imposed by theorems of this nature. Their response is one of impatience and outrage at the arrogance of mathematicians who dare tell them what they can or cannot do. The word "impossible" inspires in them an obsessive fervor. Tell them they cannot square a circle or trisect an angle and they will spend the rest of their lives trying to do it. Over a century ago the British mathematician Augustus De Morgan (whom we mentioned in the introduction) wrote the history of such "paradoxers," describing with great humor and some pathos the crotchets of these seekers of the unattainable.[2]

The pathos arises from the fact that the paradoxers do not understand and do not accept the meaning of a mathematical proof. Mathematical impossibilities arise from the meanings of mathematical objects (numbers and geometric objects) and from the definitions of the operations performed on these objects. A mathematical impossibility is not the same as a physical impossibility. It is impossible for there to be 20 eggs in a dozen, because a dozen is *defined* to be the number 12. The physical appearance of the egg has nothing to do with it. Similarly, it is impossible for an inch to contain 3.00 centimeters, because an inch is *defined* to be exactly 2.54 centimeters long. It doesn't matter what kind of ruler you use to measure it.

Similarly, the accepted value of *pi* cannot be anything other than the transcendental value we know, because *pi* is defined to be the ratio of the circumference to the diameter of a circle on a plane surface. A circle being what it is, the value of *pi* cannot be anything else. This is what the paradoxers of the world do not understand.

A tragi-comedy footnote to the story of the circle-squarers is the case of a physician, Edwin J. Goodman, M.D., who, in 1897, convinced the Indiana House of Representatives to pass a bill denouncing the conventional methods of calculating *pi,* and legislating his own paradoxical way of computing that number. Fortunately, a mathematician named C. A. Waldo happened to be in the state capitol, heard of this impending catastrophe, and convinced the state senate to postpone consideration of the bill indefinitely.[3] It is a salutary tale; legislatures of all kinds should refrain from legislating matters that are in the domain of nature.

This would all be trivial except for the fact that the effort to decide what is possible and what is impossible in mathematics has had a great effect on the history of mathematics. It has led toward the development of entire branches of mathematics, and it has improved the rigor of proof.

And for us it provides another example that demonstrates the mythology of "nothing is impossible." Anybody who says that nothing is impossible, and means it literally, is following the path of De Morgan's paradoxers.

Physical Impossibilities

A machine that purports to generate more energy than is fed into it is commonly called a "perpetual motion machine." If you had a machine like this, you could feed part of the output energy back to the input and so keep the machine running forever. A device of this kind represents an absolute impossibility in the true sense of the word. It is an impossibility that is both physical and fundamental, because it arises from the very structure of the universe. An impossibility of this kind is different from the mathematical impossibilities of the previous section. Mathematical impossibilities are based upon the creation of the human imagination and depend only on abstract concepts. Physical impossibilities are based upon the reality of the universe and so must be established by experiment and observation.

Since the basis of reality lies in the behavior of the particles composing all matter, the fundamental laws of nature are revealed by observing these particles. What we find, as a result of this observation, is that the fundamental particles behave in very specific ways governed by laws called "symmetry principles." During the past century the truly basic research in physics has focused on the nature of these symmetries —their description, their properties, and their connections with the laws of physics that describe how stars, baseballs, and even humans behave. (If you think that the laws of physics do not describe how humans behave, try to make a human behave in a way *not* allowed by the laws of physics.)

Symmetries are commonplace in nature and in art. An abstract painting composed entirely of uniform concentric circles has rotational symmetry, because if you rotate it on its axis, it does not change its appearance. On the other hand, if the painting is circular but has a design that is repeated every sixty degrees, then a more restrictive kind of symmetry is present: the design is symmetrical with respect to a rotation of sixty degrees. Snowflakes possess this kind of symmetry.

In physics one of the most important symmetry principles is time symmetry, which governs all of the four fundamental interactions between particles. The meaning of time symmetry is simply this: *the forces by which the fundamental particles interact are all independent of time.* That is, if you were to measure the force between an electron and a proton, the result would not depend on when you did the measurement or when you set your clock. The strength and direction of this force depends only on the properties of the particles and on spatial geometry: the distance between the particles, their velocity, their spins, their electric charges, etc.

What is the consequence of time symmetry? Why is it important to us? To answer these questions we must make a small diversion to mention Noether's theorem. Emmy Noether was a German mathematician whose work on symmetry principles at the beginning of the twentieth century paved the way for much important work in particle and quantum physics. (Despite that, it was only with great reluctance that the faculty at the University of Göttingen allowed her to become a colleague—without pay.[4] Being a professor in an important university does not free one of the prejudices of the period.)

Noether's theorem states that every physical symmetry is related to a physical quantity whose value remains constant while other things might be changing around it. (For example, when two billiard balls collide on a flat table, the total momentum and the total kinetic energy of the two balls are the same after the collision as they were before the collision, regardless of their speed or direction of motion.) Whenever something stays invariant in the midst of change, we suspect a conservation law at work. Thus it is that each physical symmetry is related to one of the conservation laws.[5]

The conservation law related to time symmetry is conservation of energy: no interaction between fundamental particles can result in a net gain or loss of energy. Regardless of how these particles move about or how they interact with each other, the *total energy* of the system must remain unchanged. (This includes the intrinsic energy related to the mass of the particles: $E = mc^2$.) If this law is true for all the particles of which a machine is made, then it must apply to the machine as a whole. There is no exception to conservation of energy. No matter how many particles are involved, or how complex the device or system, nothing changes the basic law.

The fact that conservation of energy applies to an entire machine, even though verified for the individual particles, is an example of reductionism at work. While many people have an objection to the concept of reductionism, its application in a situation like this is indisputable. In Myth (p. 33), I argued that this is true because conservation of energy is a bottom-up rule. More will be said about this under Myths 13 and 15.

As we have seen, conservation of energy is not only a theoretical idea. It has been verified to an extraordinarily high degree of precision by experiments of the most fundamental kind. These are the experiments that assure us the physical symmetries are more than aesthetic ideals: they are the way the universe is built. There was a time, earlier in this century, when it was fashionable for physicists to get carried away and say things like "Nature loves symmetry" in order to justify a pleasing theory. But it is important from a realist's point of view to keep in mind that only humans (and other animals) are able to love anything. Nature doesn't love anything. It has no feeling. It behaves in a symmetrical manner because symmetries are built into its structure. There isn't any other way that it can behave. It is the job of physicists to discover the nature of the symmetries and to determine the particular circumstances under which each symmetry applies.

Conservation of energy is the justification for the refusal of the U.S. Patent Office to consider applications for perpetual motion machines. Conservation of energy is also the basis for the outrage of individuals who hate it when others tell them what they can do and what they cannot do. In spite of the total failure of every perpetual motion machine in history, there are still people trying to make money out of devices that purport to put out more energy than is put into them. I have, for example, an engineering report dated October 26, 1987 from a test laboratory (Air Techniques, Inc.) which claims that the heat output of a "Perkins Furnace" carries off more energy than is provided by the electrical power going into it. Reports such as this merely demonstrate that precise thermal energy measurements in dynamic situations are notoriously difficult. (This fact has had special significance in the recent wave of headline-making "cold fusion" claims, and has undoubtedly been responsible for much of the confusion.[6]) In spite of the fact that perpetual motion machines do not really work when put to the test of generating useful work, it is still possible to make money out of them by selling stock to innocent backers. There is no law of conservation of money or credulity.

While relatively few people really believe in the possibility of perpetual motion machines, a large number of people—probably more than half the population—believe in one kind of psychic phenomenon or another. These beliefs persist even though ongoing efforts to understand the mechanisms of the brain point in a diametrically opposite direction. Current research being carried out by our best neuroscientists is uniformly based on realistic concepts—on the premise that nerves are made of normal molecules, and that information is transmitted along nerves by electrochemical pulses. There is no modern student of cognitive science who believes in "psychic energy," or who believes that "mind" is a separate entity existing outside the workings of the nervous system, or who believes that psychic phenomena are more than artifacts produced in a normal way by those workings.

Psychic phenomena—telepathy, clairvoyance, remote viewing, pre-cognition—all depend on the transmission of information from a distant source through space into the brain by some unspecified means. Regardless of how the information travels, it must trigger the motion of electrons and ions within the nerves if it is to generate a thought. This activity requires the injection of energy into the nervous system. Even though most of the energy may come from internal sources—from the metabolism of food—the initial impulse must come from the outside. The eyes, ears, and other sensory organs obtain this energy by well-known physical processes. In telepathy—transfer of information from one brain to another—the transmitting brain must convert electromagnetic energy into some unspecified form to be hurled into the space between the sender and the receiver. Psychics and parapsychologists do not even attempt to suggest ways for this to happen. They seem happily uninterested in, and incurious about, the need to conserve energy.

Even more significant from the physical point of view is the fact that any kind of energy sent through space must spread out as it goes along. As it spreads out, the energy density (the energy per square centimeter) must follow the inverse square law. That is, every time the distance doubles, the energy density must decrease fourfold. Even laser light, commonly thought to travel in a perfectly focused beam, spreads out as it goes. It simply has less spread than the broadcast light from a bare bulb, or even a searchlight beam. But after it travels far enough, the laser beam eventually obeys the inverse-square law. (This spreading out is required by the Heisenberg uncertainty principle, which tells us that each photon in the beam is going to travel in a slightly different direction.)

The basic reason for the inverse-square law is conservation of energy. If a light beam covers an area of one square meter after it travels a distance of one kilometer, by the time it has gone two kilometers it has spread out to cover four square meters, but the same amount of energy is spread out over that fourfold area. We know it is the same amount because there can be no change in the total amount of energy in the beam (unless it is lost by absorption in the medium through which it travels). Therefore, the energy density at two kilometers is only one-fourth the amount at one kilometer. Since it is the energy density that determines the brightness of the beam, the brightness must follow the same law.

The same must be true for any kind of message or signal. We expect that in all circumstances the signal strength must get weaker as the receiver gets farther away from the transmitter, and we observe this with every kind of physical communication in existence. Yet parapsychologists appear to believe that distance has no influence on the results of their experiments. (As noted under Myth 1, the astronaut Edgar Mitchell conducted an ESP experiment while he was orbiting the moon and his partner was on the earth. His paper describing the experiment was published in the *Journal of Parapsychology* without comment as to possible violation of natural law.)

Robert Jahn (Dean Emeritus of the School of Engineering at Princeton University) claims to have observed through a long series of experiments the influence of consciousness on the behavior of electronic noise generators. He lists as one of the questions to be answered by his experiments: "To what extent do the results depend on the separation of the operator from the machine?"[7] While his experimental results on this point are ambiguous, he doesn't find it at all strange to raise the question. His basic posture is not physical, but idealistic and mystical.

This attitude is expressed when he analyzes his experiments. He begins with the influence of the experimenter's consciousness on the output of a noise generator (a diode). He looks for small changes in the number of positive pulses relative to the number of negative pulses. It is like affecting the head-to-tail ratio of a series of coin flips by directing thoughts at the coin. Having established to his satisfaction that there is a real effect, he goes on to address the subject of "the focus of the interaction between the consciousness of the operator and the machine." He asks; "Is the physical behavior of the electronic noise source being affected, and if so in what way? Or, is the effect possibly more systemic, manifesting itself in the output data in an anomalous statistical form, *without alteration of any specific physical process?*" (emphasis mine). Thereupon he proceeds to replace the diode noise source with an electronic random number generator, and finds that he gets the same telepathic influence on the numbers as he did with the diode.

Clearly this experimenter is operating from an idealistic point of view. The physical processes going on within the random number generator mean nothing to him. In his theory the psychic effect works on the statistics, not on the object generating the statistics. This is not physics, nor is it reality. From a realist's point of view, when a computer generates a series of random numbers, numerous components within it (transistors, resistors, capacitors, etc.) are involved in producing the final result. You can't change the output without changing the actions taking place inside the circuit. "Statistics" is not a physical thing. It is an abstraction invented by mathematicians to help organize information about observable events. (This reification of statistics is reminiscent of the "probability wave" concept used by some of the early quantum theorists to whom probability was a physical thing capable of propagation in the form of a wave.)

The fact that parapsychologists are not terribly concerned about the mechanism by which one mind influences another mind should not be too surprising. The average viewer rarely stops to ask how the picture gets into the TV receiver. Ask owners of microwave ovens how the food gets cooked, and they will probably tell you that the cooking takes place from the inside out. How the energy gets into the center without influencing the outer layers is a matter of indifference. Yet the physical evidence is there for everybody to see. When a container of soup is defrosted, the outside layer melts first, leaving the interior frozen. The microwave radiation is a transmitter of electromagnetic energy, and this energy is deposited in the first layer of absorbing material that it encounters.

If ESP phenomena are to have realistic, physical explanations, we would expect them to follow similar rules.

More Physical Impossibilities

Physical impossibilities stem from a number of causes. Nature simply will not cooperate with any action that violates any of the several laws of denial we have already discussed. In the previous section, I invoked one of these laws—conservation of energy—to show that nobody is ever going to build any form of perpetual motion machine, and to argue that a message cannot be sent through space without it getting fainter with distance.

Just as powerful is Newton's second law of motion, modified when necessary by quantum mechanics and relativity. This rule is essentially the definition of "force" in physics. It requires any object acted on by a force to accelerate in the direction of that force. This law gives us a prescription to calculate the amount of acceleration; knowing the acceleration we can determine where an object will be at the end of a given time. Newton's second law is a good, general purpose law used to calculate the motion of projectiles, spaceships, and planets—that is, anything not too small and not traveling too fast.

While Newton's second law is not primarily thought of as a law of denial, we can cast it into a negative form by saying: "No object can remain at rest when it is acted on by an unbalanced force." (By an unbalanced force we mean a force that is not counteracted by an equal force in the opposite direction.) This simple rule makes an effective defense against some of the most cherished mythological beliefs.

Consider an Unidentified Flying Object (UFO) resting motionless high up in the atmosphere, as portrayed in any number of films. The first question that springs into my mind is: what holds the UFO up in the sky while it is being pulled down by the earth's gravitational field? If it has mass, it must be attracted by the earth. And it cannot float motionless unless another force nullifies or counterbalances the earth's gravity. Without another force, the UFO must fall.

Stories about sightings of UFOs never deal with this fundamental question. Even if the UFO is scooting along at a high horizontal speed, something must hold it up, must keep it from accelerating downward. It doesn't have any wings or helicopter rotors. It shows no signs of rockets. (And rockets couldn't hold enough fuel to last more than a few minutes anyway.) So what holds it up?

Some apologists for UFOs point to something called "antigravity." While antigravity, first employed by H. G. Wells, has been a staple of science fiction for a long time, it is necessary to keep firmly in mind that antigravity is found only in fiction. It does not exist in reality.

The reason is simple: gravity is always an attraction between pairs of masses. It is never a repulsion. It is not like the electrical force, which comes in two varieties. As a result, you cannot arrange masses in any configuration that causes them to do anything except attract each other. In addition, you cannot arrange masses in any manner that diminishes the normal attraction between any two objects. There is no gravity shield, no antigravity force generator. True, the general theory of relativity predicts certain small effects due to rotational motion, but nothing that would emulate antigravity.

In spite of these discouragements, experimenters at Tohoku University in Japan have claimed to measure a reduction in the weight of a rapidly spinning wheel.[8] Their gyroscope, mounted on a vertical axis, was weighed on a chemical balance. A 175-gram wheel rotating at a rate of 13,000 revolutions per minute achieved a weight reduction of 11 milligrams. This gives rise to the thought that perhaps if you spin the wheel fast enough, you can nullify the force of gravity. (If you can spin it fast enough without the wheel flying apart by centrifugal force, that is!) Unfortunately for UFO devotees, the experiments have been repeated and it was found that nothing in the way of antigravity could be observed.[9] The second group of experimenters (at the Bureau International des Poids et Mesures in Sèvres, near Paris) blamed the results of the first group on thermal effects due to friction, and on the way the gyroscope interacted with the balance. Whatever the reason, the initial anomalous result could not be repeated. It is a cautionary tale: be skeptical about reports of experiments which violate theory that has been thoroughly verified.

We return to the question: what holds UFOs up? Some UFO supporters cite magnetic fields. "Magnetic fields" has become a popular buzzword used by people who don't know the difference between a magnetic field and a gravitational field. The fact is that even if a UFO did generate a strong magnetic field, there would have to be another magnetic field on the ground to push against in order for the UFO to remain aloft. (Remember that all forces occur as a result of pushes or pulls between pairs of objects.) Furthermore, we know how to calculate forces produced by magnetic fields, and in this case the numbers do not compute. Magnetic fields are good for levitating trains a few inches above the tracks, but not for maintaining a massive spaceship thousands of feet in the air. Finally, if UFOs had very strong magnetic fields connected with them, these fields would generate currents in any power line they flew over, and if they did not blow out half the fuses in the country, they would at any rate be easily detectable.

The conclusion is this: no physical process exists that is able to keep a UFO supported motionless in the air for an extended period of time. This is a strong conclusion. It is not qualified by disclaimers such as "no *known* physical process." Now, whenever we come to this point in the discussion, UFO enthusiasts assure us that advanced civilizations on distant planets will have the use of forces that we are not yet aware of. Therefore we will have more to say about this matter under Myth 6. (A disclaimer: None of this argument is to deny the possibility, or even the probability, that there are civilizations on dis-

tant planets. I simply do not believe that UFOs reported in the press
are vehicles from these distant civilizations.)

The principle of relativity, discovered by Albert Einstein in 1905,
is a general law that allows us to prove the impossibility of a number
of actions. One impossibility is this: no material object or any form
of energy capable of carrying a message can travel faster than the speed
of light. Therefore, nobody is going to build a ship that travels to the
nearest star in less than a year's time. The reason is that it takes light
itself four years to reach the nearest star, and that's as fast as anything can go.

The same rule applies to messages or signals of any kind. The concept of instantaneous or faster-than-light (FTL) communication is strictly
wishful thinking and fantasy. It is easily shown that if it were possible
to send a message faster than light (in relativistic spacetime), then under
certain circumstances this message would travel backwards in time,
causing unfortunate time-travel paradoxes.[10] (The special circumstance
is that the message is sent from a spaceship traveling rapidly at a speed
less than the speed of light.) For example, knowing that a certain
catastrophe was to take place on a distant planet, you could send an
FTL message that would travel backwards in time to that planet, warning them of the impending disaster. The people on that planet would
then prevent the catastrophe. But if the catastrophe does not take place,
it would not be necessary to send the warning. But without the warning
the catastrophe would take place. And so on, and so on. Clearly it
is not possible for the catastrophe to both happen and not happen.
(At least not on the same branch of time, if it is true that spacetime
might be branched.) Whenever a paradox such as this arises in physics,
it gives us a strong hint that something is wrong. (These same paradoxes arise in connection with time-travel.)

Many writers have pointed out that the workings of quantum
mechanics allow elementary particles to communicate with each other
instantaneously and suggest that this "quantum strangeness" may make
it possible for humans to do the same. While quantum strangeness does
exist, and is truly strange, a mechanism of this kind does not enable
us to send messages or information faster than light. The reason for
this is that all the *observable* effects of quantum communication are
related to measurements made on pairs of particles traveling away from
each other and separated by a large distance. In the experiments done

within the past decade, a pair of photons traveling in opposite directions create electrical signals when they strike a pair of widely separated detectors. The instantaneous transfer of information has to do with the fact that one detector seems to know immediately the state of the distant photon. However, before a human can know anything about this correlation, the electrical signals must be transmitted through long wires to an electronic circuit (located somewhere between the two detectors) that indicates when both photons are being detected simultaneously. The electrical signals must travel more slowly than the speed of light; therefore, there is no way for any message observable on a human level to travel instantaneously from one end of the experiment to the other.

What do we make of this peculiar situation? How can we have something that travels faster than light but cannot be used to transmit messages faster than light? Interestingly, this is not an isolated occurrence; nor is it a new one. Several other analogous situations can be found in classical physics. These all involve types of waves (or other phenomena) which under certain conditions travel through space faster than the speed of light. This is a circumstance we ordinarily think of as forbidden. However, with a deeper understanding of relativity, we realize that there is no rule about steady, unchanging waves traveling faster than the speed of light. What is forbidden is the transmission of any kind of energy or information faster than the speed of light.

For example, the bright dot on the face of a cathode ray tube may travel back and forth faster than the speed of light. However, this bright spot is not a material object. It is simply the intersection of the electron beam with the face of the tube. It has no mass or energy, and you can't send a signal from one side of the tube to the other by means of this dot. Any signal carried by the electron beam must travel outward from the beam source to the tube face.

Two kind of waves that travel faster than the speed of light are (1) electromagnetic waves in a waveguide and (2) certain types of plasma waves in an ionized gas. The distance between adjacent crests (the wavelength) of these waves is greater than it would be in empty space, and when the frequency is multiplied by the wavelength, the product represents a speed called a "phase velocity," and this speed happens to be greater than the speed of light. Note that the phase velocity is never measured directly; it is simply inferred from the wavelength and frequency. In order to measure the speed of a wave, we must modulate it in some way, just as we would to send a message. When we calculate or measure the speed with which the modulation travels (called the group velocity), we find that it is always slower than the speed of light, as required by the principle of relativity.

The reason for going into such detail on this subject is to show that there is no need to become mystical about things that travel faster than light, for these are ordinary physical occurrences. There are many phenomena that travel faster than light, but in no case is it possible to send signals at this speed. And that is what counts.

Another fantasy beloved of science fiction devotees is that of time travel. To go into the past or the future is certainly desirable, and many are the schemes that have been devised for this purpose.

Curiously, there is no specific law of physics that says time travel is impossible. It simply goes against our notion of causality. When event A causes event B, somehow event B always takes place just a bit *after* event A. We never see event B taking place *before* event A. Neither do we see event B materializing far into the future. There is a continuity in nature that stems from the fact that all events take place because of interactions between pairs of particles. The cause of the interaction between atom 1 and atom 2 is an electromagnetic field. If atom 1 approaches atom 2, atom 2 cannot feel an effect until the change in the field has arrived at atom 2. The delay time depends only on the distance between atoms 1 and 2 and on the speed of light. The thing that is somewhat baffling is why the action always travels from the past to the future, because the fundamental interaction between the two atoms doesn't care about the difference between past and future. The question of why time always goes in one direction is one of the fundamental problems of science and philosophy.

Nevertheless, scientists occasionally become intrigued by the possibility of rearranging the order of events. It is clear that for time travel to take place there must be a way to bypass the normal way of making things happen, and to form a physical connection between one point in spacetime and another point. The press has recently reported some theorizing about a scheme to realize time travel by sending the traveler through a wormhole arranged in such a way that he emerges in the past or the future. A wormhole is something like a black hole, but it connects two points in spacetime, so that if you go in here, you come out there. According to this theory, it is possible for "there" to be at some point in the past. However, the theory does not specify how you arrange a wormhole to be where you want it to be. How does one manipulate a wormhole? If it is anything like a black hole, the task would be akin to placing the sun where you want to put it. What kind of force do you use? Furthermore, the theory does not address the fact that the earth is traveling through space, so that if you arrange the wormhole exit to be at another point in time, you have to put it exactly at the point in space where the earth is going to be (or was). Otherwise you end up floating in interstellar space. Science fiction writers have known about this for at least fifty years. I can admit that it is not impossible to place the near end of the wormhole where you want it, but fixing the location of the far end would appear to be a far more difficult task.

Perhaps the most compelling argument about time travel is the generation of paradoxes. The most common one is this: if you were able to travel into the past, you could then arrange to murder your grandmother. But then you would not have been born. But then your grandmother would not have been murdered, and therefore you would have been born! However, it does not seem possible to both exist and not exist at the same time. Ergo, time travel is impossible.

The common science fiction solution to this dilemma is the notion of branching universes. If you go into the past and kill your grandmother, you set up a fork in the stream of time. There is one branch where you are born and exist, and another branch where you are not born and so don't exist. After you kill your grandmother, you are on the branch where you will not exist when it is time for you to be born. Therefore, you can never get back to your starting point, and whatever changes you make in the world of the past can never reach the people who saw you begin your journey. In other words, nothing you do can make an observable change in your original time-branch. You simply disappear from sight. Thus the time-branch theory makes time travel a bit less impossible, but on the other hand, it also makes it impossible to go to another time and then come back where you started. Time travel is thus irreversible (if it is at all possible).

You realize, of course, that all of this is pure speculation. What we are doing is piling dubious hypotheses on top of other dubious hypotheses: If you can find a wormhole and manipulate it (nobody has even seen one yet), and if you can pass through it unscathed (nothing can go through a black hole without being stripped down to elementary particles), and if you can find your way back to earth after coming out of the wormhole, and if the wormhole ejects you into the past or future—then you have traveled through time.

For the same reasons we may doubt the possibility of precognition —foreseeing the future. Mystics and psychics think of precognition as a kind of "reaching out" and "knowing" what is going to happen, as though all time and space is one—as though distance in spacetime is meaningless. However, from a realistic physical point of view, we must maintain a model in which "knowing" arises from information entering the brain from outside. (Knowledge can also arise from information generated within the brain, as when you solve a mathematical problem. But knowledge about the outer world must be verified by information entering the brain.) There is no physical way for information to travel from the future to the present. In addition to the difficulties already mentioned relating to time travel, we must also be able to identify the physical carrier of the information and the source of energy required. These are the same problems discussed in the previous section under telepathy.

The same comments hold for information reaching the mind from the past through channels other than the normal physical forces. "Channelling" and reincarnation are simply not concepts included within the framework of science.

To summarize: we have identified a number of actions which are forbidden by the laws of physics:

a. It is impossible to build any device that generates energy from nothing.

b. No existing force is able to levitate human beings, or to hold a vehicle such as a UFO suspended high in the air (aside from mundane mechanical devices such as helicopters).

c. It is impossible to send information from one brain to another without the transmission of a physical form of energy. Because of this fact, it is also impossible to send a telepathic message that does not get weaker with distance.

d. For these same reasons, we must consider impossible all phenomena which involve perception of the outer world without a physical source of information-carrying energy entering the brain from the outside.

e. Conservation of energy also dooms hopes for producing physical effects on objects outside the body by projecting thoughts at them (psychokinesis). If you are trying to move an object even as small as a single electron, you must explain how the energy gets from your brain to that particle, and you must elucidate the way that energy causes the electron to move according to your wishes.

f. Instantaneous or faster-than-light transmission of objects, energy, or information must be regarded as vain ambitions.

g. Efforts to bypass causality by time travel are likewise doomed to failure.

I am well aware that many—perhaps most—people are going to find fault with my list of impossibilities. I was forcibly reminded of this recently during a lively discussion with a friend who was of the opinion that "anything is possible." It happened when I attempted to explain to him that we know some things for sure, and that this knowledge allows us to state with confidence that some things are indeed impossible. At this point he turned around and said to me: "Oh, you're talking about things we know about. I'm talking about things we don't know about!" This extraordinary response brought my mind to a halt. How can things we don't know about contradict things we do know?

After lengthy cogitation I realized what the man was talking about. He was talking about miracles. If something we don't know about can occasionally interfere with the laws of the universe and cause an event that is outside the domain of science, then we have a miracle. And with miracles anything is possible.

We arrive at an impasse. How can science deal with events that do not follow the rules of science?

The answer is simple: first prove that miracles actually take place. Then we can worry about the mechanism for their occurrence.

This is essentially what workers in parapsychology are doing. They are trying to prove the reality of miracles. The importance of this work is not to be underestimated. If miracles are real, science must rethink all its ideas.

However, while parapsychology has spent the past century trying to prove miracles with no results that anyone outside the parapsychology community can agree upon, physics has spent the past century building a colossal structure of knowledge. The learning curve is clearly in favor of physics.

NOTES

1. E. Kasner and J. Newman, *Mathematics and the Imagination* (New York: Simon and Schuster, 1940. Second paperback, 1965), p. 69.

2. A. De Morgan, *A Budget of Paradoxes* (Chicago: Open Court Publishing Co., 1915). (Originally published in Great Britain in 1872).

3. P. Beckmann, *A History of Pi* (New York: St. Martin's Press, 1971), p. 174.

4. H. R. Pagels, *Perfect Symmetry* (New York: Simon & Schuster, 1985), p. 189.

5. M. A. Rothman, *A Physicist's Guide to Skepticism* (Buffalo, N.Y.: Prometheus Books, 1988), p. 96; M. A. Rothman, "Conservation Laws and Symmetry," in *The Encyclopedia of Physics,* ed. R. M. Besancon (New York: Van Nostrand Reinhold, 3rd ed., 1985), p. 222.

6. G. M. Miskelly, et al., "Analysis of the Published Calorimetric Evidence for Electrochemical Fusion of Deuterium in Palladium," *Science* (10 November 1989): p. 793.

7. R. G. Jahn and B. J. Dunne, *Margins of Reality* (New York: Harcourt, Brace, Jovanovich, 1987), p. 103.

8. H. Hayasaka and S. Takeuchi, "Anomalous Weight Reduction on a Gyroscope's Right Rotation around the Vertical Axis on the Earth," *Physical Review Letters* (18 December 1989): p. 2701.

9. T. J. Quinn and A. Picard, "The Mass of Spinning Rotors: No Dependence on Speed or Sense of Rotation," *Nature* (22 February 1990): p. 732.

10. M. A. Rothman, *A Physicist's Guide to Skepticism,* Appendix.

MYTH 4

"Whatever we think we know now is likely to be overturned in the future."

The Invariance of Reality

If we know nothing for a certainty, then any theories which scientists claim to be true now may turn out to be false in the future.

This kind of argument is put to good, if not logical, use by many science fiction enthusiasts. If you suggest that stories about faster-than-light travel to distant stars are pure fantasy, their stock answer is: "Everything we thought we knew in the past has been overturned, so how do we know the theories of today will not be replaced by different theories in the future?"

According to the idea that theories are but temporary abstractions, the principle of relativity now in fashion will undoubtedly turn out to be wrong in the future. Somehow a way will be found to get around the part of the theory that says you can't travel faster than light. Look at history. Heavier-than-air flight was once believed to be impossible. See how silly those unbelievers turned out to be. Therefore, scientists are just as wrong when they say faster-than-light travel is impossible. Although argument-by-analogy is indefensible in any kind of science, I have heard this kind of logic proposed by some well-known science fiction writers whose reputations are based on writing "hard science."

The same logic is used to support the notions of anti-gravity, time travel, ESP, telekinesis, and other staples of the science fiction repertoire. What makes this argument invalid is the fact that it is based upon a myth. The idea that all theories are temporary is simply not true, even though it is believed by a great many people. The reason is, as we have shown, that we do know some things for a certainty.

Even though the history of science is filled with examples of scientists changing their minds about the nature of things, does it necessarily follow that they will continue to change their minds about all presently accepted ideas? The truth is that many of the theories of the past were primitive, tentative, and either incorrect or incomplete. When they changed, they changed to theories that were more correct and more complete. This change took place because as laboratory techniques improved, experimental evidence got better.

Undoubtedly there are fads and fashions in science, and there have been some cases where theories were believed regardless of the evidence. (The recent case of "cold fusion" is a perfect example of a dubious belief held by a group of scientists while the experimental evidence was either negative, marginal, or contradictory—while the physical theories needed to explain the evidence were nonexistent.) Nevertheless, there are no examples in twentieth-century physics of a theory enduring for more than a brief period of time in the absence of good empirical evidence. There are plenty of non-established theories lurking in the wings, waiting for the validating evidence to establish them. But for those of our presently accepted ideas that are well validated and established, there is little chance that they are going to change in any substantial manner.

Some historians and philosophers of science try to claim that scientific knowledge is nothing more than the consensus of the scientific community. According to this notion, scientists need only change their consensus and knowledge changes accordingly. Even in the "soft" sciences such as psychology and sociology, I wonder if "knowledge" based on consensus and subject to the whims of opinion should be dignified by calling it knowledge. It should better be called a collection of unvalidated hypotheses.

In the physical sciences the consensus theory of knowledge simply does not take the facts into account. Much of the conventional wisdom of nineteenth-century physics consisted of a variety of conjectures believed to be true by a number of scientists and accepted on faith by others without close examination. It did not conform to our present standards of "knowledge." It was, for example, accepted that light was composed of waves in a mysterious medium called "the ether." Heat was taken to be an excess of a fluid called "caloric." "Phlogiston" was responsible for combustion, "animal magnetism" was responsible for hypnosis, and "creationism" was the accepted account of human development. Even Einstein briefly accepted the tradition of the ether before his discoveries of 1905 turned it into a superfluous notion. To reduce scientific knowledge to "consensus" ignores the contribution of empirical evidence—evidence based on observation.

Empirical evidence was raised to its present status towards the end of the nineteenth century when two developments in scientific methodology produced a qualitative change in the way we look at nature.

1. Scientists, following the lead of Gustave Kirchoff and Ernst Mach, began to discard the use of unobservable entities such as phlogiston and caloric. Instead they adopted the attitude that scientists should not ask "Why?" but rather "How?" In so doing, they discarded metaphysics and progressed to the position that implies science can only deal with observable events behaving according to specific natural laws.[1] In essence they were making use of *Occam's razor,* a rule of thumb set down by William of Occam in the fourteenth century: "Entities are not to be multiplied without necessity." In other words, don't invent things whose existence produces no measurable or observable consequences. For a period of time many scientists, including Mach, carried this attitude to such an extreme that they refused to believe in the existence of molecules, even though the molecular theory did predict numerous measurable consequences, e.g., the relationship between temperature and pressure in a gas.

Nowadays the rules have been softened: we do not insist that everything be *directly* observable. All we require is that the things we talk about produce effects that can be observed, and that there exist a set of rules (correspondence rules) connecting the nonobservables to the observables.[2] (In the theory of gases, such a correspondence rule would be the relationship between the kinetic energy of the molecules and the temperature of the gas.) Indeed, even the concept of "observable" has changed. We now think that an atom has been observed when it has produced an image of some kind on a computer monitor screen, or a black spot on a photographic film, even though numerous processes have taken place between the detection of the atom and the production of its picture.

While atoms and molecules can now make their presence known on an individual basis, occasionally it is necessary to assume the existence of an intrinsically invisible entity in order to explain the things we see. For example, we now find it necessary to postulate the existence of six kinds of quarks in order to explain the properties of all the observed particles. These quarks do not exist separately and so cannot be observed even indirectly, but their use allows us to calculate mathematically the properties of the particles that we do see—for example, their masses. Furthermore, quark theory includes an explanation for the fact that quarks cannot be observed: the binding force between a pair of quarks increases as you try to separate them. If you separate them too far, you must put so much energy into the system that new particles are created. Thus, instead of seeing the separate quarks, you see more protons, neutrons, and mesons of various kinds.

Consequently we are cautioned that in creating a new theory (such as quark theory), we must make sure that the theory is both necessary and sufficient. For a theory to be necessary, there must be no other theory that explains the same data. For the theory to be sufficient, it must be able to predict everything that is observed (within the range of the theory), and explain all the experimental results, without exception. Following these rules increases our confidence about the permanency of the theory.

2. The second change in methodology at the end of the nineteenth century was the explosive growth of instrumentation, leading to the discovery of a universe invisible to the naked eye. This universe contained structures on a minute scale: elementary particles. It also contained structures on an immense scale: galaxies and clusters. Both the instrumentation and the identification of the fundamental entities of the universe led to the adoption of philosophical realism as a guide to pragmatic science. With pragmatism, science became much more than a matter of theory spinning. Realism requires us to start with the premise that the universe is filled with real objects: galaxies, stars, planets, living organisms, molecules, atoms, and particles. The task of science then becomes directed toward determining the nature of these objects and the nature of the laws that determine their behavior. Theories are not mere inventions of the mind. While invention plays a role in interpreting the signals from our instruments, *the theories are about things that exist in the real world and the nature of the theories are determined*

by the structure of the real world. In order for a scientific theory to be believed, there must be some reason to believe it.

It is true that quantum mechanics has softened the hard outlines of realism. When we deal with single elementary particles we find that they do not follow all the strict laws of classical mechanics. The laws of quantum mechanics give predictions concerning the chances of some event taking place; they do not always say exactly what event is going to take place. Two colliding atoms do not act like billiard balls. During the collision, they do things which are incomprehensible according to the ideas of classical mechanics. Nevertheless there is order to the uncertainty of their actions. Even though we cannot predict where the colliding atoms will travel after the collision, we can be sure that the collision will not change the total momentum and energy of the system. The fundamental laws are still safe.

Two colliding billiard balls, on the other hand, do act like billiard balls. The reason is that when we get millions of particles taking part in one action, the uncertainties concerning their average motion become so infinitesimally small that they may be ignored. Thus realism applies to human-level activities.

Most bewildering to those steeped in classical thinking is the fact that some of the properties of observed particles depend on the way the observation is arranged. If you set up an interferometer experiment to measure wavelengths of light, you observe waves. If you set up a photoelectric cell to look at the same light beam, you don't observe waves, but you observe particles (photons) instead. More important, it is impossible to set up an experiment that observes both properties simultaneously. In the interferometer experiment, it must be remembered, the things actually observed are the photons emerging from the interferometer; the waves within the interferometer determine how many of the photons land in a given region on the film. But the wave action took place before the particles were detected.

So what is it we are observing? Is it a particle or a wave? Our language is incapable of expressing the hidden reality. But whatever it is, something real has been observed; the instrument extracts from the real thing the property it is able to measure. The words we use to describe our observations are high-level interpretations of the interaction between the real thing and our instruments.

Realism (as opposed to idealism) has been the engine driving science during the past century. All progress in science has been an outcome of realism. By contrast, there has been no accumulated knowledge based on idealistic theories. This difference in the outcome of realistic and idealistic theories is not often mentioned explicitly in textbooks or in the popular press, but it is implicit in all contemporary scientific literature. It also becomes clear that intellectual conflicts such as the current controversy between evolution and creationism stem from fundamental differences in outlook between realists and idealists.

Realism assumes that the entities observed by the instruments of science are objects with measurable properties. These objects are the subject-matter of science. Knowing that we are dealing with real objects supports our belief that the universe, once created, does not change without cause. Everything we learn about the nature of the stars that existed billions of years ago teaches us that nothing fundamental has changed about the properties of atoms since that time—the kinds of atoms and the kinds of forces that existed at the beginning of our galaxy still exist today. This observation leads us to believe that particles and forces do not change their nature without a reason.

When we examine the discarded beliefs of the past, we find that these beliefs uniformly violated the rules of realism. Belief in the ether was never more than a speculation unsupported by the kind of evidence we now consider necessary. It was simply unsubstantiated opinion. With the advent of Maxwell's equations and Einstein's relativity, light became recognized as a phenomenon related to electromagnetic fields. What advantages have electromagnetic fields over the ether concept? The advantages lie in the fact that electromagnetic fields can be precisely and operationally defined. This means that you define them by describing how to measure them. Having done that, you can then go ahead and perform those measurements with appropriate instruments. (Every radio and TV receiver is essentially a measuring device for electromagnetic fields.)

By contrast, there was nothing about the ether that could be detected or measured.

However, that's not the end of it. Planck and Einstein's quantum theory (1900–1905) demonstrated that certain observable effects (the photoelectric effect, for example) could only be explained if we assumed light to be composed of particles, i.e., the photons. The photon, as it turned out later, provided a more fundamental explanation of light than electromagnetic fields. A single photon does not come equipped with an electric field or with a direction of polarization. Yet when many photons pass through the proper instruments, we obtain effects that we ascribe to an oscillating electric field, and the direction of this electric field is the direction of polarization of the wave. Electromagnetic waves are seen to be a high level abstraction or interpretation found to be useful when many photons act in unison. (In exactly the same manner, we find that water waves are the result of countless water molecules acting together under the influence of gravity.)

Early quantum theory made the photon the carrier of electromagnetic energy from one atom to another. Light, x-rays, gamma rays, even radio waves were composed of photons. But as the twentieth century progressed, quantum theory evolved, endowing the photon with even deeper meaning. The science of quantum electrodynamics made the photon not only the carrier of energy, but also made it the carrier of the electromagnetic force that determines how charged particles interact with each other. The electrical repulsion between two electrons is the observed consequence resulting from the interchange of many photons between these two electrons. Quantum electrodynamics for the first time gave us a *mechanism* for one of the fundamental interactions. Shortly thereafter the standard model of particle physics was developed —a theory in which the photon was only one of four kinds of *bosons,* each of which was the carrier of one of the four kinds of known forces. In each case, the concept of "force" became a high-level abstraction, representing the end result of multitudes of phantom bosons being exchanged among the particles.

In less than a hundred years physics experienced a dynamic evolution of the theories that claimed to explain the behavior of electromagnetic fields: the classical electromagnetism of Maxwell and Schroedinger's quantum theory evolved into quantum electrodynamics, which in turn became the gauge theories that described the interactions of fundamental particles. No wonder the general public can believe that science constantly changes, and that no theory is permanent. The state of flux characteristic of physical theories is enough to bewilder anyone.

However, to keep our orientation as we thread our way through this maze, we must follow a simple rule: *Look for those things that stay constant while everything else is changing.* Among all the changes that have taken place in theories of electromagnetism during the current century, a number of facts and ideas have remained unchanged.

These are two of the most important:

1. The concept of the photon has remained basically invariant even as the theories surrounding it have become more sophisticated. The central feature of the photon is the relationship between the energy of the photon and the frequency of the wave that it represents: the photon energy is equal to Planck's constant multiplied by the frequency. This relationship has remained unchanged since Max Planck first suggested it in 1900.

2. The importance of the speed of light (the unique speed of all photons) as one of the universal constants of nature has remained unchallenged. The most refined experiments have demonstrated that no matter how it is measured, no matter how the source or detector of the light is moving, the speed of light is an invariant—unchanging—quantity. That makes the velocity of light different from all other velocities. The velocity of everything else depends on the motion of the observer. Not so for light, which is why its constant speed is the basis of the principle of relativity. Of course, as with any other measured quantity, there is some small uncertainty to this statement; there are some error bars on the graphs. However, the limits of error have been pushed to such an infinitesimal amount that the invariance of the speed of light is now accepted as an established piece of knowledge for all theoretical as well as practical purposes. It is part of the foundation of all physics. (All of special relativity, incidentally, can be derived from the observation that the speed of light is constant.[3])

The result of this knowledge has revolutionized our concepts of space and time, and has had a profound effect on our ways of measuring both space and time. A hundred years ago our standard of length measurement was the distance between two scratches on a platinum rod. Then, with improvements in instrumentation, the standard of length became, in 1960, the wavelength of a certain line in the spectrum of krypton-86. Now, however, our precision in measuring the speed of light has overcome the accuracy with which wavelengths can be measured. Therefore, in October of 1983, the velocity of light was adopted as the standard of velocity and was *defined* to be exactly 299,792,458 meters per second. The decimal points were arbitrarily lopped off.

It is a legitimate question to ask how physicists can get away with such drastic action. Is it not like defining *pi* to be exactly 3.14? Not at all. The number *pi* is a mathematical quantity and cannot be tampered with. The speed of light, on the other hand, is a physical quantity —a velocity—defined to be the standard by which all other velocities are compared. A physical standard can be defined to be anything you want, as long as it has the proper dimensionality. Choosing the speed of light to be an exact standard replaces a standard of length by a standard of velocity. This standard of velocity has no uncertainty. All of the measurement uncertainties have been handed over to the value of the time standard, the second. The meter is now a secondary standard, defined in terms of the standard velocity and the standard second of time.[4]

The development of quantum theory and particle physics during the twentieth century shows us that science—at least the basic physical sciences—does not lurch from one theory to another, demolishing all past ideas as it goes. Instead it evolves, discarding the ideas that fail, while it absorbs those that work into new concepts that work even better.

Invariant Knowledge

If the nature of reality is invariant, then at least some of our knowledge of that reality should be invariant. Otherwise, the effort of the scientific enterprise to determine the nature of the universe has come to naught. To counter the myth that "whatever we know now is likely to be overturned in the future," we must examine the enormous amount of knowledge available to us and ask ourselves: what are the chances that all of this knowledge is going to change? Next, we try to identify the types of knowledge that we believe are permanent. Some of these questions have already been examined under the rubric of Myth 1: "Nothing is known for sure."

To begin with, we know that the earth is not flat, and that it is not the center of the solar system. The solar system is composed of planets moving about under the control of the mutual gravitational attraction between pairs of planets and satellites. It would require skepticism of the most extreme sort to think that this kind of knowledge is going to change in the future.

In physics it is not too difficult to identify permanent knowledge. The idea that matter is composed of atoms and molecules, while controversial only a century ago, is now the basis of all science. Modern instrumentation not only allows us to observe the structure of individual atoms, it also allows us to manipulate single atoms so as to place them where we desire. On April 5, 1990, the Associated Press displayed in newspapers all over the country a photograph of the characters "IBM" formed by placing 35 xenon atoms onto a nickel crystal. Each xenon atom was put into position by an electric probe which served the double purpose of imaging the crystal surface and dragging the individual atoms into place so as to form the English characters. Knowledge of the properties of these atoms and of the forces which allow us to manipulate them is secure and permanent knowledge.

How do we really know that atomic structures and forces remain constant with time? Our knowledge comes from astronomy. We are able to analyze the spectra of the light coming from galaxies billions of light years away. This light originated in the atoms of stars that existed billions of years ago. The patterns of wavelengths observed in these stars is identical to the patterns we find in the light from the sun (allowing for Doppler shifting due to the motion of the distant stars). This indicates to us that the photons that come to us, the atoms, the electromagnetic forces, and the laws describing how the atoms interact with these forces have not changed since the universe began.

Occasionally a physicist does question seriously the constancy of the natural laws. There was a time when it was legitimate to ask whether the strength of the gravitational force relative to the electric force is constant or has very slowly changed over time. Such a change could be observed as an alteration in the rate of clocks operating by electrical forces (the frequency of a laser beam) relative to clocks operating by gravitational forces (the earth revolving around the sun). However, within the accuracy of the experiments no such change has been observed. Of course, we cannot use this result to claim that there is absolutely no change. There could be changes so slow that it would take more than the present age of the universe to observe them. This means that for all practical purposes there has been no change, and this is the meaning we will use when we speak of "no change" in the paragraphs that follow.

Our failure to find temporal changes in the fundamental constants leads us to believe that once a particle has been detected and its properties measured, these properties are not going to change. Once verified, a natural law is not going to change.

However, the words used to describe a theory may change from time to time, even though this theory may be completely verified. Quantum electrodynamics—the quantum theory of electromagnetic fields and their interaction with elementary particles—is the most successful theory in physics. It has been experimentally verified with exquisite precision and has been used to predict many effects previously unknown. There is no reason to expect a change in the observable consequences predicted by this theory. However, new ways of putting the theory into words or into mathematics may arise. The underlying reality of nature is not going to change, and the things we observe are not going to change. However, our descriptions of this reality may change as we acquire more insight and as our instruments allow us to see finer details in the observed phenomena.

For example, in "classical" quantum theory it was conventional to describe a particle as being represented by a wave packet and to speak of the "reduction or collapse of the wave packet" upon detection of the particle. However, there are versions of quantum theory in which wave packets are not even mentioned. Furthermore, it can be shown that the idea of wave packet reduction leads to paradoxical consequences.[5] The wave packet is simply a human invention which helps visualize the reality described by certain equations. Other ways of describing quantum reality are likely to be invented in the future. It is possible that the actual reality may never be truly known, but the observable consequences predicted by the theory are the invariants that make up permanent knowledge.

Another example of the ways in which we use different words to describe the same reality are the disparate gravitational theories that exist. Newtonian theory speaks of fields of gravitational force that spread out in all directions from each mass, affecting the behavior of all other masses in space. Einstein's theory of general relativity, on the other hand, does away with the concept of force fields, describing gravitation as caused by the curvatures of space-time generated by the masses embedded in it. A quantum theory of gravitation describes the gravitational interaction as the result of the exchange of gravitons between pairs of masses. These are all different ways of describing the same thing. All of these agree with the observations to a high degree of precision, but the predictions of relativity are better than the predictions of Newton. Furthermore, relativity predicts some observable effects that are not revealed by Newtonian theory (the precession of the orbit of Mercury, the bending of light paths around the sun, the relation between gravitation and the Coriolis force, etc.). To be accepted, any future theories must agree with the observed motions of the planets to better than one part out of a billion. It must also explain black holes, the evolution of the universe, and other matters that are presently handled by relativity. In other words, any new theory of gravitation must encompass the old theory, do all of its tasks better, and also explain any new observations that come along. Otherwise, there is no need for a new theory.

A new theory may give results very similar to those of an old theory even though it is based on a profoundly different set of concepts. When we calculate the motion of objects moving not too rapidly, the results obtained by using the special theory of relativity are hardly distinguishable from those computed from Newton's laws of motion. This is no coincidence, for it was necessary for relativity to give the same results because it had to explain the same data. From this fact arises a minor myth of science perpetrated by some writers who say that "the special theory of relativity reduces to Newtonian mechanics at low velocities."

While it is true that many of the equations of relativity reduce to Newton's equations when the velocity approaches zero, we must keep in mind that the conceptual basis of Einstein's theory is totally different from that of Newton. In relativity, the only kind of motion that means anything is the motion of one object relative to another object. There is no meaning to "absolute motion" in relativistic mechanics, while absolute motion is the very basis of Newton's mechanics. Thus, special relativity not only encompasses Newtonian physics, it provides a new conceptual framework from which emerges the relativity of time, the equivalence of mass and energy, and the relationship between electric and magnetic fields. It is fortunate for Newton that the speed of light is such a large velocity. Had it been as small as, say, the speed of sound, Newton would have had a much more difficult time in establishing the laws of motion. The world would have appeared very different, and Newton would have had to be Einstein.

In the second half of the nineteenth century, it was possible for the most eminent scientists of the time, e.g., Ernst Mach and Friedrich Ostwald, to disbelieve that atoms and molecules existed. However, knowledge has changed—and now, in the second half of the twentieth century, it is not possible for any scientist not to believe in the existence of atoms and molecules. This change is irreversible, for the network of evidence is too intricately interwoven for the structure to collapse. Once we know that atoms exist, it is impossible to turn back the clock and show that atoms do not exist. Once they are discovered, the discovery cannot be undone.

Much of the confusion concerning the permanency or changeability of knowledge stems from the fact that knowledge exists in a variety of forms:

a. *Existence of objects.* A realistic theory requires that things in the universe do exist. In order for us to believe in the existence of any object, we must observe it, identify it, and measure its properties. When scores of scientists all over the world observe the same object and find identical properties after employing different measuring techniques, then they can be confident of its existence. Atoms can now be photographed on an individual basis. Knowledge of their existence is not going to change. Details may change. Atoms may behave in ways we have not yet thought of. (High temperature superconductivity has yet to be explained, for example.) The electron may have a structure at present only suspected. But the existence of the electron is not going to change. Its mass, charge, and other measured properties are not going to change either.

The number of known particles tends to increase with time as more kinds of particles are observed. New theories tend to organize this variety of particles in different ways. Quark theory did not eliminate the large number of mesons and short-lived baryons that had been observed. They still exist, but they are no longer considered to be *fundamental* particles; they are made up of quarks, just as the hundred-odd atoms are combinations of electrons and nuclei.

b. *Existence and nature of forces.* The four forces of the standard model have been identified and most of their properties are known. There are still details to be completed. The nature of the weak force has many uncertainties, and measurements of the numbers of neutrinos coming to us from the center of the sun indicate there is something we don't understand about the nature of neutrinos. So the standard model is not complete. In addition, gravitation has yet to be quantized, and new theories being studied propose to show that the four forces are fundamentally a single force. Important for our purposes is to understand that the existence of these forces is not going to change. Nor are the known properties and symmetries of these forces going to change.

c. *Quantitative measures* are indeed subject to some change. We cannot say that the value of the charge on the electron or the gravitational constant is absolutely constant. But, as I pointed out earlier, any changes to occur in the future must be small enough to be imbedded within the present error. (The error bars themselves may have to be corrected, as in the case of measurements of the speed of light. Early experimenters optimistically underestimated the size of the error bars, with the result that the speed of light kept shrinking as the measurements became more accurate.)

d. *Theories,* being human descriptions of reality, can change as better observations are explained by more detailed theories. But, as we have seen above, newer and future theories cannot contradict the observations that have already been made, unless those observations are grossly in error. No laws of motion can ignore the fact that the speed of light is a constant. It is impossible for a future theory of gravitation to show that objects repel each other by a gravitational force—unless that repulsion is so weak that it has not yet been observed. (Indeed, there have been efforts to measure a repulsive component of the gravitational force. The results were negative; premature claims of positive results were caused by errors in interpreting the data.)

At this point in the argument, somebody is sure to say, "You can't rule out the possibility that a miracle might change the gravitational constant some time in the future. You are simply assuming that just because things appear to be invariant now, they are going to remain the same forever."

Indeed, this is the very argument made by the creationists when they claim that rates of radioactive decay suddenly changed some time during the past 10,000 years, causing measurements of the age of fossils to be totally false. The purpose of this claim is to prove that the earth is less than 10,000 years old, as required by creationist theory.

However, that kind of argument removes the discussion from the domain of science. In science there are no miracles. Things happen because they are required to by the laws governing the universe. Within the existing universe there is no cause for a change in the rates of radioactive decay. Creationists hypothesize such a sudden arbitrary change because they need it to prove their creationist theory. But this is a circular argument. The creationists start with the existence of fossils, shown to be millions of years old by radioactive dating, under the assumption that the isotope lifetimes have been constant during all that time. However, they want to prove the earth is less than 10,000 years old. Therefore, they assume that a change in radioactive decay

rates took place shortly after the creation. The only evidence for this change is the fact that it allows them to "explain" the million-year-old fossils. The argument assumes what it claims to prove. The creationists start out by assuming the earth is 10,000 years old, and then make up a story about radioactive decay to "prove" that this assumption is valid. However, they have no independent empirical evidence for this change in decay rates, no ideas about the amount of change, how it varied for different isotopes, and no numbers that allow quantitative calculations. There is no evidence that radioactive decay rates have actually changed at any time. (Indeed, their hypothesis ignores the fact that detailed events within the atomic nucleus determine the rates of radioactive decay.)

By claiming that certain things happened 10,000 years ago to make the earth appear 10,000 years old, you can "prove" anything you want. You can claim that fossils were placed in the crust of the earth by the maker of the universe just to make it appear there was a biological evolution. You can even claim the earth was created by God ten years ago, complete with a population and memories of past eras. There is no way to test such a hypothesis and no way to prove it false. Thus the hypothesis is not falsifiable and is therefore not an empirical scientific theory based on observed evidence. It is a theory based on miracles.

We cannot say *a priori* that miracles are impossible. But for us to believe that these miracles exist, we need unambiguous empirical evidence. And so far such evidence has not been presented. Miracles—in the sense of events not easily explained—certainly happen within the mind. But miracles which violate the verified laws of nature are harder to come by.

NOTES

1. R. Carnap, *Philosophical Foundations of Physics,* ed. M. Gardner (New York: Basic Books, 1966), p. 12.

2. Ibid., p. 233.

3. E. F. Taylor and J. A. Wheeler, *Spacetime Physics* (San Francisco: W. H. Freeman & Co., 1966), chap. 1.

4. E. R. Cohen, "Fundamental Constants," in *The Encyclopedia of Physics,* 3rd ed., ed. R. M. Besancon (New York: Van Nostrand Reinhold), p. 227.

5. M. A. Rothman, *A Physicist's Guide to Skepticism* (Buffalo, N.Y.: Prometheus Books, 1988), p. 68.

MYTH 5

"Advanced civilizations of the future will have the use of forces unknown to us at present."

Imagining the Future

Classical science fiction—fiction that has some basis in legitimate science —is an exercise in imagining possible futures. The science fiction of the 1930s generated a number of prophecies which, now that we live in what was then the future, have come true: nuclear energy, television, computers, music synthesizers, worldwide electronic communications, space travel, genetic engineering, etc. One important prediction remains in limbo: the idea that we can learn how to build a computer that can think like a human being and is aware of its own existence (a true robot). The jury is still out on this one; we cannot prove that the idea is impossible, but we are a long way from implementing it. As for sending men and women to Mars, no physical impossibility deters us; only the difficulty of obtaining sufficient funds stands in the way.

There remains one stubborn class of prophecies which have not come true, and, in the opinion of many, never will. These prophecies have to do with faster-than-light (FTL) travel, time travel, antigravity, interstellar wars, and an entire spectrum of psi phenomena activated by powers of the mind—extrasensory perception (ESP), telekinesis, telepathy, precognition, and so on.

Hard-core science fiction enthusiasts and parapsychologists still believe that some of this latter group may yet be realized. The prospect appeals to the legendary sense of wonder that suffuses the mind of the science fiction reader. To these people, all options should be kept open, or else the universe becomes cold, static, and closed. Nevertheless, considering what scientists know about the laws of nature, I must reluctantly conclude that none of these exciting fantasies are going to materialize.

The reasons for coming to this unhappy judgment have been covered in the previous chapters. In brief: the structure of spacetime makes faster-than-light travel an unattainable goal. Conservation of energy makes psi phenomena just as impossible as perpetual motion, and for the same reasons. These are the inescapable conclusions of our present knowledge.

However, people determined not to lose their dreams of the future do their best to escape from these conclusions. After all the scientific arguments are presented to them, their invariable response is that "advanced civilizations of the future will have more knowledge than we have now, and so will have the use of forces more powerful than any we now control."

Notice there are two parts to this statement. First, that advances in knowledge will result in the discovery of forces presently unknown. Second, that these new forces will be powerful enough to accomplish tasks that we now consider impossible: to reverse the force of gravity, to establish conditions necessary for FTL travel or time travel, to move and transform entire planets, and to transmit messages instantaneously.

Yet, when we think seriously about the nature of forces, when we ask ourselves where forces come from and how they work, we come to a dismaying conclusion: nobody in the future is going to discover any new and exotic forces—forces strong enough to produce important visible effects felt over long distances. That is to say, using the jargon of modern physics, there exist in the universe no strong, long-range forces that are not already recognized. And if they do not exist now, they will not exist in the future.

Let us see why I say this.

Do We Know All the Forces?

First, let us be clear about what is meant by a force in physics. In ordinary discourse, when one is "forced" to do something against one's will, some powerful influence has been brought to bear against the victim: threats of violence, blackmail, poverty, and so on. An armed force is a group of men and machines capable of impressing its will upon an enemy and causing a change in its behavior. A force, then, is an instigator of change.

This is true in physics also, but there the concept of force has a precise, quantitative meaning: it is something that causes an object or system of objects to change its state of existence. According to Newtonian mechanics, a force is considered to be anything that changes an object's state of motion. In fact, Newton's second law of motion is essentially the definition of force: it says the amount of force acting on an object is equal to the object's mass multiplied by its acceleration.

In engineering mechanics, force is impersonal. It is exerted by one object onto something else. The earth pulls you down, and if you are unsupported, you fall, your acceleration being given by Newton's second law. However, in modern physics, forces always arise as the result of a *mutual* interaction between two objects. The split between the exerter and the exertee is purely a human convenience. In actual fact, you always pull up on the earth just as much as the earth pulls down on you, incredible though that may seem. But that is the meaning of Newton's third law.

In modern particle physics, the concept of force is generalized into a somewhat abstract notion. Instead of using the word "force" when dealing with elementary particles, we speak of the *interaction* between the two particles. Most notable is the fact that a force is *always* an interaction between two objects. There is no such thing as a "disembodied" force, or a force acting only on one object. Furthermore, nobody has discovered a force that requires an interaction between three or more objects.

The most important new feature of modern physics is that an interaction may have consequences more complex than the mere acceleration of a particle. When two energetic electrons interact, they do more than just bounce off each other. Totally new particles may be generated. Even at reasonably low energies, photons may be created by this collision.

At higher energies neutrons, protons, or mesons may spray off in all directions. An interaction may also cause the disappearance of existing particles. The interaction between a low-energy electron and a low-energy positron invariably results in the annihilation of the two particles and the creation of a pair of photons in their place. There is no law that says the total number of particles taking place in a reaction must remain unchanged. Only the total energy of the system must remain constant, as well as the momentum, angular momentum, charge number, baryon number, and a few other invariants.

As we noted previously, the standard model of particle physics recognizes only four kinds of forces or interactions existing in nature. More advanced theories—Grand Unified Theory and superstring theory—lead us to believe that these four interactions are part of a single unified force, but for our present purpose it is convenient to speak of four separate forces.

These four interactions account for every kind of activity that occurs—either on the subatomic level or on the cosmological level. Each interaction has a unique set of characteristics. These are: the kinds of particles interacting with each other, the kind of particle that carries (or mediates) the interaction, the interaction strength, and the range of the force—that is, the way its strength varies with distance. These characteristics are outlined in Table 1.

The most familiar force, and the one first identified by Newton, is the gravitational interaction. This force manifests itself as an attraction between all masses. The other classical force is the electromagnetic interaction. Though familiar, the electromagnetic interaction is quite complicated. The electric part of the force acts as an attraction or repulsion between electric charges. The magnetic part of the force arises from the motion of electric charges and acts at right angles to the direction of motion. In addition, whenever charges are accelerated, electromagnetic waves are generated, i.e., photons are emitted. A most important difference between the gravitational and electromagnetic forces arises from the fact that there is only one kind of mass, but two kinds of electric charges. Thus, the gravitational force can never be other than attractive, while the electric force can be either an attraction or a repulsion.

The weak nuclear force acts between electrons, positrons, muons, and neutrinos. It is responsible for certain kinds of radioactive decays, and is of importance only in nuclear and particle physics. The strong nuclear force manifests itself as an attraction between pairs of quarks, and it is responsible for the structure of neutrons and protons, as well as for holding together the neutrons and protons within each atomic nucleus. Thus, the strong force is responsible for nuclear structure.

The interaction strength varies over an enormous range. (This strength is computed by assuming two particles at a distance of 10^{-13} cm from each other, center to center, and comparing the magnitude of each interaction with the strength of the strong nuclear force, seen to be the most powerful.) Scanning the table, we see that the strong and the electromagnetic forces are much more potent than the other two. The gravitational force is considerably weaker than the electromagnetic force, and only gives the impression of being strong because the stars and planets have such great masses. Actually, the electrical repulsion between two protons is 10^{36} times greater than the gravitational attraction between the same two particles. Therefore, the electrical force predominates in any happenings at the atomic and molecular level. However, when planets and stars interact, the attractive and repulsive electrical forces neutralize each other because an equal number of positive and negative charges are present. The gravitational force is then left to take over.

The interaction range is seen to come in two types. The long-range forces follow the inverse square law. Even though the force diminishes in strength as the two objects get farther apart, it never completely vanishes. The short-range forces, on the other hand, make themselves felt only within distances smaller than the diameter of an atomic nucleus (about 10^{-13} cm). Because there are strong forces and weak forces, as well as short-range and long-range forces, we find it convenient to arrange the four forces into the two-by-two array shown in Table 2.

This array brings out the extraordinary fact that there are only two long-range forces observed in nature: the gravitational and the electromagnetic. The other two forces are short-range and only make themselves felt within the atomic nucleus, or in other circumstances (such as within the centers of stars) where two particles can get very close together. In other words, the gravitational and electromagnetic forces are responsible for everything that happens in the universe outside the atomic nuclei. Gravity controls stellar and planetary motion, as well as the motion of footballs and acrobats. The electromagnetic force is responsible for molecular structure, the behavior of solids, and for life itself. The electromagnetic force is seen to be unique. It is the only force that is both strong and long-range. Thus it is the only force that can be the cause of important effects on a human level.

We now ask the two questions that are central to this chapter:

1. What is the probability that new forces exist which we have not yet discovered?

2. What is the probability that advanced civilizations in the future will find useful forces that we do not already have?

These questions can be answered in a simple way: If there were any long-range forces in existence other than the two already known, and if their effects were strong enough to be readily observable, then we would have observed them already. The fact is that all the observed behaviors of the known particles are explainable in terms of the four interactions of the standard model.

This would appear to be a pretentious claim. There are plenty of things in the world that we cannot as yet explain (high-temperature superconductivity, for example). But nobody expects that explanations for new phenomena on a molecular or solid state level will require the introduction of radically new forces and particles. It is true that there may be new and unknown short-range forces making themselves felt at very short distances or at very high energies, but these can have no relevance to the science-fictional manifestations previously mentioned: antigravity, transmission of telepathic messages, faster-than-light travel, etc. These all involve events happening outside the nucleus and so require long-range forces for their implementation.

In addition, none of this argument denies that there may exist new long-range forces that are so weak their effects have not yet been observed. Presently under investigation are a variety of hypothetical gravitational forces. One of the new theories involves a repulsive type of gravitational force that acts only at distances less than a few hundred meters. Another proposed force affects protons differently than neutrons, so that objects made of different isotopes would fall at different rates. But these new postulated forces are a hundred times weaker than the standard gravitational force, which is already the weakest of the four forces. The observable effects attributed to these new, hypothetical forces are so minute that experimental results are still controversial. At the time of writing, it appears that positive results claimed by some experimenters are due to errors in data analysis (false assumptions concerning the distribution of the masses within the earth).

Our conclusion, then, about the existence of exotic new forces: there are no unobserved long-range forces in nature with interactions strong enough to cause the extraordinary properties claimed for them by enthusiasts of the extraordinary.

The only question to be answered then is: can new and unknown forces be created in the future with the aid of techniques we know nothing about? In answering this question we must keep in mind the fundamental concept of particle physics: all forces arise as a result of interactions between pairs of particles. The known forces are interactions between known particles. In order to invent new forces, we would have to invent new particles that do not exist in the present universe.

There is nothing in physics to prevent the creation of new particles. Physicists have been doing this regularly since early in the twentieth century. Every time they build a new particle accelerator and collide electrons or protons at higher and higher energies, they create particles never seen before. However, all the new particles fit into the scheme of the standard model (see chapter 2), and the astronomical evidence assures us that these are the same kinds of particles that existed at early moments of the universe when particle energies were very high, and the conditions in the universe at large were similar to those that exist momentarily during particle collisions in the target zone of a high-energy accelerator.

The crucial feature of the standard model—as far as this discussion is concerned—is the fact that all the particles of the standard model can be classified into three families, each containing four particles. And the latest experimental results (described under Myth 2, p. 59) indicate that these three families include all the particles that can exist under the ordinary conditions of living matter.

This extraordinary result puts a damper on speculation about advanced civilizations of the future creating new particles and new forces with which to carry out the wonderful advances hoped for by science fiction enthusiasts.

An obvious rejoinder to this argument is to point out that the past century has witnessed an exponential growth in knowledge of sciences and uses of technology, and then to ask why these advances should diminish in rate or quality in the future. Surely the exponential curve is going to continue its upward trajectory.

The answer to this question is found by examining the kinds of advances that have been made during the past century, and to note that they are not all of the same kind. We start by defining two categories of scientific progress. These categories seem arbitrary at first, but we will see what makes them important.

1. Knowledge associated with the gravitational interaction and with the strong and weak nuclear forces:

The gravitational interaction is concerned entirely with the nature of the universe at large: astronomy, astrophysics, and cosmology are all intimately tied in with twentieth-century discoveries concerning the nature of gravity (general relativity).

Discoveries of new particles, the development of nuclear physics and the standard model of particle physics are all connected with the strong and weak nuclear forces.

2. Knowledge associated with the electromagnetic interaction:

This includes: special relativity, quantum theory, atomic physics, the physics of solids, liquids, gases, and plasmas, chemistry, electronics, molecular biology, genetics, and all other aspects of biology.

There is a great difference between the above two categories. Everything in category 1 is of the greatest theoretical and philosophical importance. Astronomical and cosmological knowledge is needed to know our place in the universe. But it has very little to do with our efforts to control nature, to build mechanisms, to direct forces, to communicate information. Nuclear physics plays a role in understanding stellar mechanics, in generating power, in medical technology. Indirectly, the topics in category 1 have played a strong role as an incentive for the development of instruments used in scientific research. But the development of these instruments belongs to category 2.

Category 2 is responsible for the technological advances that have an impact on the affairs of individual human beings. Our mastery of solid-state physics, coupled with the development of information theory, has fueled the computer revolution. Our understanding of organic molecules has caused biomedical science to explode. All of the knowledge in these categories is associated with the behavior of electrons and protons, their response to electromagnetic fields, and to nothing else.

The basic reason for this dichotomy is the fact that humans are able to manipulate electromagnetic fields. Gravitational and nuclear fields, on the other hand, are beyond our ability to control. The control of electromagnetism begins within the human brain: signals consisting of electron and ion currents cause the contraction of muscles in other parts of the body. With these muscle motions we can turn generator cranks so that a coil of wire rotates in a magnetic field and causes an electric current to flow. This current can cause a motor to rotate far away; it can generate electromagnetic waves in a transmitter.

All of this happens because ordinary matter is made of electrons and protons and neutrons, and the structure of this matter is such that in some kinds of matter electrons are free to flow, and in other kinds of matter they are not so free to flow. By understanding how to make use of the structure of matter, and by manipulating this structure (as in the manufacture of semiconductors), humans are able to produce devices that do not exist freely in nature.

Due to the fact that there is only one kind of mass (as opposed to two kinds of electric charges) we are unable to manipulate gravitational fields. We cannot make a gravitational generator or insulator. Nothing that we do has an effect on the planets or stars as astronomical objects. (Note that our ability to change the temperature of the earth's surface by the greenhouse effect is basically an electromagnetic phenomenon.)

Similarly, there is nothing we can do to manipulate the nuclear forces. Even when we manage to make slight alterations in rates of radioactive decay by varying the pressure exerted on the source, this effect is caused by electromagnetic fields reaching into the atomic nucleus.

The point of this digression is that the technological advances of the past century have been mainly due to our ability to manipulate electromagnetic fields. Nuclear and gravitational fields we take as they come. We use only the particles and forces that we find in nature. Therefore, when we speak of creating new forces in the future, it is irrelevant to point to our advances of the past century as an indicator of future growth. It is a false analogy. Past technological growth has had little or no relevance to the creation of new forces. In fact, the opposite is true: the more we learn about particle physics, the more we are forced to recognize that there is a limit to the number of different kinds of particles and forces that can exist.

The conclusion to which we are forced—unsatisfactory as it might be to many—is that we cannot depend on the discovery of new and radically different kinds of forces in the future to help us go faster than light, to hold vehicles suspended in midair, to make objects move by directing thoughts at them, or to transmit messages telepathically. We must make do with the forces that exist.

TABLE 1

CHARACTERISTICS OF THE FOUR FORCES

Force	Strength	Range	Carrier
Gravity	10^{-38}	inverse square	graviton
Electromagnetism	10^{-2}	inverse square	photon
Weak nuclear	10^{-13}	10^{-16} cm	intermediate boson
Strong nuclear	1	10^{-13} cm	gluon

TABLE 2

RANGE-STRENGTH CHART

	Long-range	Short-range
Strong	Electromagnetic	Strong nuclear
Weak	Gravitational	Weak nuclear

MYTH 6

"Advanced civilizations on other planets possess great forces unavailable to us on earth."

The Uniformity of the Universe

Let us drop in on the middle of a typical argument about the reality of UFOs. The skeptic has just finished saying that he doubts the strange things people see in the sky are vehicles from a planet circling a distant star.

Believer: "You think we're the only intelligent beings in the universe?"

Skeptic: "It's not a question of us being the only ones. There probably are plenty of inhabited planets elsewhere in space. The question is whether those things you think you see in the sky came here from another star."

Believer: "What else can they be? Lots of people have seen them all over the world."

Skeptic: (Sarcastically.) "Sure. So they come here faster than light, and they hang up there in the sky humming away like a banshee."

Believer: "Why not? You think that's impossible?"

Skeptic: "Yes, it's impossible. There's no way to nullify the force of gravity, and there's no way to go faster than the speed of light."

Believer: "You don't know what advanced civilizations might find. Those people from other stars are way ahead of us scientifically. They have machines and forces that can do things we can't imagine."

And there we have the myth: the idea that advanced civilizations on other planets will have the use of forces unavailable to us here on earth and will be able to do things that we consider impossible.

This belief is false for very much the same reasons stated in Myth 5. An advanced civilization is bound by the same physical laws and limitations as the most primitive civilization. The advanced civilization can do more with what it has, but eventually it bumps up against the same physical limitations that control all actions in our less advanced society. The properties of spacetime prevent faster-than-light travel. The properties of the gravitational interaction prevent the making of an antigravity device. More knowledge and more technology do not change these fundamental properties of the universe.

Another question is then insinuated into the discussion: how do we know that in distant reaches of space the properties might not be different? Perhaps many light years away matter consists of different kinds of particles with greatly different properties. The people living there would be able to make use of forces totally new and strange. They would have the ability to do things impossible in our local corner of space.

We can answer that question by looking at the messages we get from those distant regions of the universe. Yes, we do get messages from stars millions of light years away, but they do not come in UFOs, and they do not come faster than light.

These messages appear in the form of light from the stars, x-ray emissions, radio waves, and the mysterious bursts of energy that we call cosmic rays. Cosmic rays are high-energy particles that inhabit the space within and between galaxies, and which reach the earth frequently enough to be detected by instruments high in the atmosphere. Instruments on the earth's surface detect mainly the products (muons, photons, etc.) formed by collisions between the cosmic ray particles and the atmospheric particles. The primary cosmic rays are known to be atomic nuclei of many kinds, ranging from hydrogen up to iron and beyond. Measurements based on the known decay of radioactive isotopes indicate that cosmic rays travel as much as 10 million light years before reaching our solar system.

The proportions of the various elements found in cosmic rays are similar to the proportions of the same elements found in the stars. This indicates that the cosmic rays started out as matter ejected with great energy from various stellar bodies. The exact mechanisms are still somewhat mysterious.

The important fact for our consideration is that the cosmic rays consist of the same kind of matter that we have in our own solar system. This is evidence that the kind of matter existing millions of light years away from earth is the same as the kind of matter we have here on earth.

Another kind of evidence is obtained by analyzing the visible light emitted by distant stars and galaxies. The light emitted by the hot gases in the atmospheres of these stars consists of discrete patterns of wavelengths—the lines of the spectrum. Hydrogen, the most common stellar element, emits the simplest spectrum of all—an easily identifiable set of wavelengths. When the emitting star is very far away, these wavelengths are shifted towards the red (the famous Doppler shift) because of the star's velocity away from us. However, when the spectrum is corrected for the Doppler shift, the wavelengths are identical with those emitted by our own sun. This evidence indicates to us that even galaxies billions of light years away contain the same kind of hydrogen that our own solar system contains.

Add to these facts the well-known spectrum of the black-body radiation permeating the universe, and we have the basis for the fundamental model of the universe: the idea that the entire universe originated in an explosion that took place at a single place and time somewhere between 10 and 20 billion years ago. If everything in the present universe originated from one primordial mass, then it must all consist of the same kind of matter. There is no possibility that other kinds of matter, and other kinds of forces, and other kinds of laws exist elsewhere in this universe.

Some physicists have entertained the theory that part of the matter in the universe is composed of "antimatter": positrons, antiprotons, antineutrons, etc. No evidence supporting this theory has been found, and some evidence opposing it exists. Even if it was true, this theory would not invalidate the arguments of this chapter, for antimatter is a normal part of the standard model. It does not allow for new forces or new laws.

We are left with uniformity: the cosmos is one. It is uniform in composition throughout all space. It is uniform in laws of behavior and types of forces. It is isotropic in the sense that if you look in any direction, you see the same kinds of things. There is local clumping, and there is invisible matter, but matter everywhere is fundamentally the same.

MYTH 7

"There are more things in heaven and earth, Horatio, than are dreamt of in your philosophy."

The Infinity of Facts and the Finitude of Science

There are those who achieve great satisfaction by pointing out the obvious: that scientists will never know everything there is to know. The universe is so vast; there are so many stars; there are so many facts to learn— we can't know everything. Simply cataloging all the stars would exhaust the memory storage of a major computer. If there are planets circling many stars, it would take an eternity to name all the species of all the animals living on all of these planets. And considering how many years it would take to reach galaxies billions of light years away, the need to visit all existing stars would extend the search beyond the future lifetime of the sun.

The pleasure engendered by this observation arises from knowing that there will always be an unknown to explore, that there will always be mysteries to solve. The titillation of mystery motivates the joiners of cults and the readers of esoterica. Umberto Eco vividly evokes a delicious sensation of mystery simply by declaring commonplace mathematical truths in recondite and portentious prose.[1] To many, the sense of wonder that comes from contemplating the unknown is more fun than the sense of accomplishment that comes from ripping open the barriers to knowledge and feeling that one actually understands how the universe works. For that reason, they want to feel that some of these mysteries will never be solved.

There is no arguing with the fact that the universe contains more facts than mankind could possibly organize and catalogue. However, the accumulation of facts is not science, although it is of great importance in the beginning stages of an embryonic science. "Knowing all the facts" is not the same as "knowing everything." Putting the facts into an organizational scheme and knowing the rules governing that organization are what make a science.

For example, many biologists consider it a worthy project to determine the structure of all the genes in the set of human chromosomes. This effort requires finding the sequence of the 3 billion nucleotide pairs within the DNA molecules that make up the chromosomes. By itself, this project does nothing more than determine a set of facts. By itself, it makes only the beginning of a science. However, organization and use of this knowledge will make it easier to trace the origin of genetic diseases and will make it possible to determine how each gene contributes to the function of the human body. Study of DNA structure will make it possible to trace the evolution of the human species. The genome structure is not the end product of the investigation; it is only the starting point for the research to come.

Similarly, the spectroscopists of the nineteenth century devoted a great amount of effort to cataloging the wavelengths of light given off by each element when heated to a high temperature in a gaseous state. This information was useful for identifying elements by spectroscopy, but at that early stage it was cookbookery; it was not a true science. It could not become a science until the advent of atomic physics and quantum mechanics, when it became possible to set down the rules that now enable us to calculate the wavelengths of light given off by each element. The theoretical rules are relatively brief but they contain within themselves all the information needed to calculate the emitted wavelengths. Knowledge of atomic structure and dynamics is fundamental science; the thousands of different wavelengths of light given off by the various atoms are a series of incidental, although useful, facts.

A physical theory consists of concise equations that contain within themselves a vast array of unwritten and unspoken information. To unfold this information may require the scientist to grind through a large amount of appropriate mathematics. Schroedinger's equation, for example, fits on one line of type, but thousands of solutions to the equation may be found, each of which describes the structure of a different atom or ion. How do thousands of solutions arise from a single equation? They arise because an atom can exist in numerous states with different amounts of energy or angular momentum, each corresponding to a different set of quantum numbers. A different solution is found every time a different combination of quantum numbers is entered into the equation. There is an infinity of possible solutions, but once you know how to solve the equation, the rest is trivia.

Theory, then, is a succinct language. It contains the rules which allow us to calculate what is going to happen in innumerable discrete situations. Knowledge of general rules, laws, and principles is the real content of a science. Disassociated facts, by themselves, do not make a science.

Therefore, to say that there are more facts in the universe than we can ever know is an inconsequential observation. The important question is: are there more laws, rules, and principles in the universe than we can ever know?

To phrase the question differently: *Is the number of potential laws of nature finite or is it infinite?*

If the laws of nature are infinite in number, we can look forward to a future in which no matter how many laws we learn, there will still be an infinite number of laws remaining to be discovered. On the other hand, if the number of laws is finite, then there will come a time when we can say we really know all the fundamental laws. Which is it to be?

We can look for an answer to this question by examining the physics of fundamental particles. As we have seen, the standard model of particle physics suggests that there are just three families of particles, each with four members, and that there are no more than four different ways for these particles to interact with each other. If this is true, then the fundamental rules of physics are limited in number. We need simply learn the basic laws governing the four interactions and in principle everything else follows from these basic rules. The symmetries obeyed by the four interactions determine which actions are possible and which are impossible. The nature of the particles and of the forces that guide them determines what kind of structures they will organize themselves into and determines how these structures will behave.

As of now our knowledge of the standard model is incomplete. We do not know how to calculate the mass of an electron from basic principles; we do not even know whether or not an electron has an inner structure. We do not know which particles are truly elementary and therefore cannot be divided into smaller particles. We do not know whether or not there is an infinite regress—a never-ending series of particles within particles. If there is an infinite regress, then we can never know all the laws.

It is to answer these questions that experimental particles physicists carry out their work with high energy accelerators. By forcing energetic electrons and protons to collide with each other, they hope to find new particles or to determine the inner structure of known particles.

Experiments of this kind led physicists to believe in the existence of quarks after a period of several years when the theory of quarks was tantalizing as the savior of particle physics, but nobody knew how seriously to take it. Even though quarks are never observed as separate entities, the multiplicity of mesons and hyperons observed in cosmic rays and accelerator collisions had led to the development of quark theory as a simplifying model. In quark theory all the dozens of heavy particles were considered to be combinations of just a few different kinds of quarks. Each of the possible quark combinations was associated with a specific particle—except for one combination whose particle could not be found. It was as though in a million poker games, nobody ever came up with a royal flush. Thus several years went by with an embarrassing gap in the repertoire of observed particles. Instead of there being too many particles, there were too few.

However, the theory was finally saved when, in 1964, the omega-minus baryon was observed in the laboratory. This discovery completed the set of particles which had been predicted by quark theory. The discovery of the omega-minus was the turning point that caused physicists to believe in quark theory as something more than a useful fantasy. In addition, experiments performed at the Stanford Linear Accelerator in the late 1960s and early 1970s showed that when high-energy electrons were scattered from protons within a nucleus, their scattering pattern could be explained only by picturing each proton as a mass with a grainy structure. No longer could the proton be considered as a hard, uniform lump.

For a short time there was the heady feeling that we finally knew the fundamental particles and that there was no lower level to reach for. However, it gradually became apparent that there still exist some observable phenomena that can be explained only on a level deeper than that provided by quark theory. Many physicists now believe that such a theory may be necessary to provide the unification of the four interactions, or to show the connection between particles and the "gauge bosons" that mediate the four interactions.

For this reason, we cannot rule out the existence of structure in the universe at levels below the currently established particle model. The question which cannot be answered at the present time is: does matter subdivide indefinitely, or does the subdivision stop at some point? If at some point there is an end to the regression and the smallest particles are identified, then in principle, at least, all the laws of nature can be known.

It is to answer questions such as these that particle physicists want to build expensive machines such as the superconducting supercollider. Those who object to the cost sometimes argue that a machine such as this cannot provide any knowledge useful to society. However, the questions addressed by high energy research are fundamental to philosophy as well as to physics. The subject is at least as important as theology.

Until we can obtain empirical answers to these questions, we may resort to analysis. For example, suppose we admit that matter possesses structural details down to such deep levels that we will never be able to observe or decipher all of them. This would mean that some questions will never be answered and that some types of knowledge will be forever unattainable. We then ask: does it make any difference as far as we are concerned? Do these deep sub-quark levels of physics have any effect on the capabilities of human beings?

It is true that what happens at the very bottom of the particle ladder determines the structure and properties of the particles that make up matter at the top of the ladder. But can it have an effect on anything happening outside the particles, outside the atomic nuclei? The only external effects predicted by the standard model of particle physics are those things mediated by the gravitational and electromagnetic interactions. Everything else is short-range.

This means, then, that the presence or absence of sub-quark physics is of no relevance to most of the topics that concern us in this book. Short-range forces cannot be the carriers of signals over long distances. All the manifestations of ESP—telepathy, perception at a distance, or precognition—require the transmission of energy into the brain from elsewhere. But no long-range interaction with the necessary characteristics exists in nature. And no short-range interactions will do. Transmission of information from one place to another can only come about when elementary particles interact with each other. But nowhere in particle physics can we find a force that can claim responsibility for the transmission of telepathic messages.

The point of all this discussion is that it makes no difference if we do not know all the details of the world at the bottommost levels of the quark hierarchy. As far as the theory of parapsychology is concerned, it is sufficient to know what happens several levels above, i.e., at the atomic level.

The moral: even though we may not know everything, we know enough to decide between possibility and impossibility. We know that any action is impossible if it disobeys any of the verified symmetries of nature. For this reason, no thought can be injected into your mind unless it is accompanied by enough energy to stimulate some electron and ion motion in the nervous system. Consequently, there is no necessity to believe in any of the paranormal phenomena related to ESP unless a physical means of carrying that energy from one place to another is identified. Researchers can do parapsychology experiments for the next millenium, but if they cannot supply a mechanism for the effects claimed, there is no real need to believe that the pictures created in a subject's mind originated anywhere but inside that mind.

Heavenly Mysteries

The person who raises the "more things on heaven and earth . . ." argument often seems to be more interested in heavenly mysteries than in mundane scientific matters. Keep in mind that the context of Shakespeare's words refers not to the infinitude of empirical knowledge, but rather to the existence of ghosts in the world.

The implication is that no matter how much we learn about science, there are other kinds of experiences, less quantifiable, that cannot be explained by the ordinary mechanisms of science. According to this point of view, a supernatural dimension of the world is required to account not only for the mystical phenomena of ghosts, poltergeists, reincarnation, etc., but also for the paranormal workings of ESP, telepathy, psychokinesis, and other manifestations generally subsumed under the heading of psi phenomena.

It is difficult to write coherently about this myth from the viewpoint of a scientist, since the basic premises of the mystical attitude are diametrically opposed to the scientific attitude. The differences between idealism and realism, already discussed under Myth 1, are of the same nature. The topic is an elusive one. Yet some things can be said—if only to define what questions need answering.

If you believe in ghosts, then you have to decide how the image of the ghost gets into the brain of the beholder. Does the ghost exist objectively? If so, then how does it generate photons to produce an image in the viewer's eye? From where does its energy come? Of what kind of matter is it made? If it is not made of matter, then how does it make photons? On the other hand, if the image of the ghost is strictly subjective, how does it get into the mind of the viewer? Where does it originate? Further, if the ghostly vision is manufactured within the mind, how does it differ from a hallucination?

The same considerations apply to poltergeists. Where does a poltergeist get the energy needed to toss dishes and furniture around the house? What is the nature of the coupling between the poltergeist and the thrown dish? Is it a new force, or is it one of the four natural interactions?

"But that's just the point," exclaims the mystic. "It's not one of the known forces. It's something beyond known earthly processes. The fact that ghosts and poltergeists exist proves that there are forces in the universe beyond our earthly knowledge."

This would all be very well if it was indeed proven that ghosts and poltergeists exist. However, investigations by trained observers have invariably found these phenomena to be either natural events misunderstood by the observer, or simply tricks played on the public by professional or amateur charlatans.

The same arguments apply to psi phenomena. If it was proven that psi phenomena really existed, we would have to consider seriously the possibility of forces beyond the natural. That is the real reason for spending time on parapsychology research—both as a believer and as a skeptic. But does hard proof really exist? Parapsychology experiments invariably involve small effects producing statistical fluctuations that are just on the edge of chance. To believe that telepathy, precognition, and psychokinesis have been proven by these experiments requires a real act of faith. To believe that the marginal effects claimed to date can be put to practical use requires an even higher degree of faith.

The results of a century of parapsychology research are summed up by Ray Hyman, a distinguished and skeptical student of the psychic world: ". . . whatever the reason, each generation's best cases for psi are cast aside by subsequent generations of parapyschologists and are replaced with newer, more up-to-date best cases. Not only does the evidence of psi lack replicability, but, unlike the evidence from other sciences, it is noncumulative. It is as if each new generation wipes the slate clean and begins all over again."[2]

The noncumulative nature of parapsychology research is perhaps the most important reason for maintaining skepticism towards the subject, particularly when it is contrasted with the exponential accumulation of knowledge in realistic science.

The remark is often made that "extraordinary claims require extraordinary proof." I do not think it is necessary to go that far. I would accept extraordinary claims if they were accompanied by merely ordinary proof—the kind of proof that would be accepted by a competent epidemiologist in judging medical research.

One current medical topic of interest is that of health effects caused by ordinary 60-cycle magnetic fields—the kind produced by power lines and electric blankets. We know right at the start that health risks associated with these fields must be relatively small, since the life expectancy of the population has doubled during the past century, while the use of electricity has increased many-fold. Therefore, one is looking for small statistical effects in the midst of a large background, and the use of careful epidemiological methods is required. In this case, it is at least possible to think of known physical mechanisms that might cause the claimed effects since we understand how ions are affected by alternating magnetic fields. Even with that advantage, the epidemiological studies to date offer contradictory results and must be considered as the preliminary stages of convincing research.

Similar contradictory conclusions are presented by some of the studies in nutrition periodically quoted by the media. One study says eating X lowers your cholesterol level by y percent. The next study that comes along says eating X actually does nothing to your cholesterol level. A third study says something else, and the public is left thoroughly confused. A good epidemiologist warns the public to pay no attention to preliminary studies. In dealing with small statistical effects, it is necessary to amass a large amount of reproducible statisics, often requiring years or even decades of work.

In evaluating claims of parapsychology research, we start out with an immediate disadvantage vis-a-vis work on biological effects of magnetic fields. The parapsychology work comes with no coherent theory —that is, there is no known mechanism for transferring information through space from one mind to another. Therefore, before we can believe that something is really happening in the way of telepathic communication, the evidence for it must pass rigorous epidemiological tests. The statistics must be collected according to well-known rules, there must be suitable control groups, and the deviation of the data from chance expectations must be convincing. This means that the number of tests must be sufficient in number to give reliable data.

Robert Jahn (see Myth 3, p. 81) has developed, in the course of his parapsychology research, a way of acquiring an enormous quantity of statistics in a relatively brief period of time by using the speed of modern electronics.[3] His experiments claim to demonstrate how the output of noise generators and random-number generators is influenced by the consciousness of the experimenter, causing the generators to produce a numerical output that goes a little bit higher or lower than one would expect from chance. (The effects are in the range of three standard deviations from the expected mean.) Taking advantage of the speed of a random number generator, one can simulate millions of flips in a short time.

Using my own computer, I have duplicated a portion of Jahn's experiment. A program written in BASIC, making use of its random number generator, allowed me to do 40,000,000 flips in a few hours. The results did not differ significantly from what would be expected from a pure random number generator. Either my brain does not generate enough will power to force the averages up or down, or I am one of those skeptics whose presence in the room destroys the operation of any parapsychology experiment.

Without spending a lifetime analyzing precisely what was done by each person taking part in Jahn's experiment, it is impossible to determine just how his positive results were obtained. For this reason, we will have to wait until the experiment is replicated a number of times by other people in other laboratories before we can have real confidence in its results.

In the meantime, the noncumulative knowledge obtained by parapsychology research is not sufficiently strong to convince a realistic skeptic that there are more forces in nature than those included in the standard model of particle physics.

Is There More to the Brain Than a Collection of Atoms?

A third aspect of the "there are more things on heaven and earth . . ." myth is the idea that the human brain must consist of more than a mere collection of atoms and cells operating under the ordinary laws of physics, because, in the opinion of those espousing this view, it is absurd to think that consciousness, awareness, and feeling can emerge from the operation of a computer-like device.

In previous centuries, the concepts of *elan vital* and *psychic energy* have been conjured up to set the human organism apart from non-living and non-thinking mechanisms. However, these entities have fallen into disrepute in scientific literature and are seldom mentioned in the scientific community.

Nevertheless, a number of people with reputations in the scientific world continue to feel that there must be more to a human being than a brain nestled within a body, and the literature of mind and body continues to grow.

Robert Jahn has attempted to explain how consciousness can directly influence his random number generators. His approach is unabashedly mystical when he says: ". . . far beyond its strictly intellectual capacities, consciousness possesses and utilizes mystical and metaphysical dimensions for its conceptualization and interpretation of reality."[4]

"*Reality,*" in Jahn's model, "encompassing all aspects of experience, expression, and behavior, is constituted only at the interface between consciousness and its environment. . . ." Furthermore, ". . . the sole currency of any reality is *information,* which may flow in either direction."[5] There is no hint in these words that reality may exist independently of consciousness—that the universe exists objectively regardless of human observation. The idealistic philosophy is clear from the start.

In attempting to formulate a theory of consciousness analogous to quantum mechanics, Jahn quotes Sir James Jeans, another scientist with a penchant for mysticism: "It seems at least conceivable that what is true of perceived objects may also be true of perceiving minds; just as there are wave-pictures for light and electricity, so there may be a corresponding picture for consciousness."[6]

Consciousness is presumed to exist as a wave-particle phenomenon *separate from and in addition to* the activities of the brain. The notion that consciousness emerges from brain activity is rejected. Consciousness comes first. Indeed, the possibility is entertained that consciousness enters the picture at the atomic level: "Other scientists and philosophers have pondered whether atomic structure may be characterized by its own intrinsic form of consciousness. By the definition of consciousness proposed . . . atoms and molecules would certainly qualify, for they have the capacity to exchange information with each other and with their environment and to react to these in some quasi-intelligent fashion."[7]

Roger Penrose, the Rouse Ball Professor of Mathematics at the University of Oxford, also expresses his discontent with current computer models of the mind.[8] Penrose, a very distinguished and creative scientist, must be read with serious intent. He begins by stating his reasons for believing that "strong artificial intelligence (AI) devices" cannot create the phenomena of consciousness, awareness, and feeling that characterize human beings. (A strong AI device is any kind of device that performs thinking functions by means of a well-defined sequence of operations, known as an *algorithm*.) He then goes on to claim that the currently known laws of physics are incapable of explaining consciousness and suggests that a future theory, which he calls "correct quantum gravity" (CQG), will provide an answer.

There is considerable difference between the Jahn approach and the Penrose approach. Jahn is explicitly mystical and frequently appeals to quotations by well-known scientists whose origins were within or prior to the nineteenth century, and who were known to introduce mystical ideas into their later writings. Penrose, on the other hand, makes an effort not to be mystical. His proposed additions to physics are not supernatural, but are extensions of known science.

Yet Penrose makes a number of statements that have a decidedly idealistic flavor to them. He repeatedly makes use of the concept of "Plato's world of mathematical concepts," in which mathematical ideas have an existence of their own and inhabit an ideal world which is accessible only by way of the intellect. Is he simply making a metaphor, or is he actually giving a measure of reality to Plato's world?

The answer would appear to lie in his discussion of communication between mathematicians: "When mathematicians communicate, this is made possible by each one having a *direct route to truth,* the consciousness of each being in a position to perceive mathematical truths directly, through this process of 'seeing.' . . . The mental images that each one has, when making this Platonic contact, might be rather different in each case, but communication is possible because each is directly in contact with the *same* externally existing Platonic world!"[9] Where this Platonic world is located, he does not say.

At another point, in discussing the evolution of consciousness, he asks how natural selection could have been clever enough to evolve consciousness and all the neural paraphenalia that goes with it. He says, "There seems to be something about the way that the laws of physics work, which allows natural selection to be a much more effective process than it would be with just arbitrary laws."[10] The difference between the laws of physics and "just arbitrary laws" is not made clear.

The impression I receive from these statements is that a mystic is trying to break out of the unconscious areas of Penrose's mind, but that he hides this from himself by couching his ideas in scientific form.

Perhaps the weakest link in his argument is his insistence that devices based on "strong artificial intelligence," using fixed algorithms, cannot account for human consciousness, and therefore, a new quantum gravity theory must be operating within the brain. This is in a sense a straw man, set up for the purpose of being knocked down. Is fixed algorithm AI the only kind of computer conceivable? MIT's artificial intelligence laboratory has been experimenting with tiny robots whose microprocessor brains "challenge some fundamental assumptions about the nature of reasoning and intelligence, producing startlingly life-like behavior with simple stimulus-response reflexes—and virtually none of the elaborate symbol processing used in expert systems and other traditional AI programs."[11]

The circuits used in these robots constantly alter their responses according to sensor input from an ever-changing evironment. This kind of thinking is analogous to the kind of everyday unconscious work we perform when we walk about during the day, unaware of what each individual muscle is doing. Rodney Brooks, the researcher who originated the idea, calls this way of building a robot "subsumption architecture."

Considering that the study of intelligence is essentially in its infancy, it would appear premature to pass final judgement on our ability to create an intelligent computer. Penrose seems to be saying, "I don't understand how the mind works. It seems impossible for the mind to operate according to the laws of physics that we know. Therefore, we must postulate that the mind operates on the basis of some new kind of rules, about which we know even less."

There is nothing wrong with looking for new rules. The question is, is this the most fruitful path to follow in trying to understand the mind?

The history of the past few centuries shows that scientific knowledge started its exponential growth when it started replacing idealistic ideas with realistic ideas. There has never been a scientific advance that was based on mysticism, or on vague and idealistic theories. This fact must be kept in mind when we ask ourselves whether the myth in the title of this chapter is totally false. Are there indeed more things on heaven and earth than are dreamed of in our philosophy? There are indeed new things to be discovered. But those who expect them to be of a supernatural character are playing a losing game, while those who think that subnuclear processes or quantum strangeness play a decisive role in the operation of the nervous system have yet to come up with a convincing argument.

NOTES

1. U. Eco, *Foucault's Pendulum* (New York: Harcourt Brace Jovanovich, 1989).

2. R. Hyman, "A Critical Historical Overview of Parapsychology," in *A Skeptic's Handbook of Parapsychology,* ed. P. Kurtz (Buffalo, N.Y.: Prometheus Books, 1985).

3. R. G. Jahn and B. J. Dunne, *Margins of Reality* (New York: Harcourt Brace Jovanovich, 1987).

4. Ibid., p. 37.

5. Ibid., p. 204.

6. Ibid., p. 211.

7. Ibid., p. 317.

8. R. Penrose, *The Emperor's New Mind* (New York, Oxford: Oxford University Press, 1989). (Quotations by permission of Oxford University Press.)

9. Ibid., p. 428.

10. Ibid., p. 416.

11. M. M. Waldrop, "Fast, Cheap, and out of Control," *Science* (25 May 1990): p. 959.

MYTH 8

"Scientists don't have any imagination."

Are New Ideas Always Resisted?

The genius of Albert Einstein has been obscured by a number of myths. One of these is the idea that the pioneering work of the young Einstein was scoffed at by most "establishment" scientists because of its brilliantly revolutionary nature and that, as a result, he had a hard time gaining recognition in the scientific community. The story is an effort to depict scientists as unimaginative people who always ridicule new and unfamiliar ideas, and who resist the rise of new and brilliant competitors.

True, it is easy to find a few isolated examples of physicists, both important and not-so-important, who said silly things that revealed their inability to understand relativity theory. Even the great Hendrick Lorentz, who developed the equations of relativity even before Einstein, never really accepted some of the physical consequences of these equations. The same is true of Jules Henri Poincaré, perhaps the world's most prominent mathematician, who developed all the mathematical formalisms of relativity, but failed to carry it to its logical conclusion by giving up the notion that absolute velocities can be measured relative to the "ether." Most people find it very difficult to change the foundations of their philosophies late in life.

If we ask how Einstein's career fared in the world of professional physicists, we find that whatever resistance he encountered on the road to academic acceptance was due more to politics and anti–Semitism than to physics. Einstein himself expected opposition and criticism prior to publication of his famous 1905 paper on relativity in *Annalen der Physik*. To his surprise, the immediate response to this publication was nothing other than deep silence.[1] Apparently nobody knew what to make of this new phenomenon. Finally, the quiet in the scientific community was broken by a letter from Max Planck, Germany's leading physicist, asking for clarification of some obscure points. Einstein was gratified to know that at least somebody had read the paper. The "somebody" turned out to be a most influential sponsor, and it was clearly due to Planck's interest in it that relativity quickly became an accepted topic for discussion and research.

It must be remembered that when the first relativity paper was published, Einstein—though he had just received his doctorate from the University of Zurich—was still a civil service employee in the Swiss Patent Office. There was no reason for anybody to expect a revolution in science to emerge from that small room on *Gerechtigkeitsgasse*. Nevertheless, Einstein quickly became friendly with many of the leading physicists in Switzerland and Germany, and in 1907 he won an appointment as *Privatdozent* at the University of Berne—the first step on the conventional academic ladder.[2] Two years later, Einstein was a professor at the University of Zurich. In 1912, Wilhelm Wien (the Nobel laureate for 1911) wrote a letter to Stockholm recommending that Einstein, together with Lorentz, receive the Nobel Prize in physics. In the face of this favorable reception by the professionals of physics, it's hard to accept the myth that scientists lack imagination. The revolution in the fundamental concepts of physics—from Newton to Einstein—took less than ten years to accomplish, a mere moment in history, and a very reasonable period of time compared to the acceptance times of other major intellectual revolutions.

The ironic twist to this story is the fact that during the past eighty years, the principle of relativity has gone from revolutionary to establishment science, and those unhappy with the restrictions imposed by the speed of light complain that it is the followers of Einstein who are holding back progress.

The fact of the matter is that in physics, professional physicists are almost always the initiators of new ideas. The subject has become so abstract and complex that the amateur scientist has very little chance

of formulating an idea that is both new and valid. To do so, the amateur would have to educate himself into becoming a professional. This has been done, but the number of known cases are very few.

The example of Oliver Heaviside (1850–1925) is somewhat extraordinary and provides some evidence for those who say that new and original scientific thinking is invariably scoffed at. Heaviside, handicapped by both deafness and a lack of formal education, had an uncle named Charles Wheatstone (inventor of the Wheatstone bridge) who encouraged him in his efforts at self-education. So successful was Heaviside that his work in the mathematics of electric circuits has become a standard part of modern electrical engineering. He introduced modern vector notation into the study of electromagnetic fields, originated the Heaviside operational calculus (a method of solving differential equations now absorbed into the generalized topic of mathematical transformations), and predicted the existence of a layer of ionized gas in the upper part of the atmosphere, now known as the Kennelly-Heaviside layer. Resistance to Heaviside's original ideas came from those who were unable to penetrate his novel mathematical notation, as well as from those who complained that his operational calculus lacked rigorous proof (although it happened to be right). Unfortunately for his detractors, his recognition was complete by the 1930s, and texts on mathematical physics universally taught Heaviside's vector analysis and operational calculus. The outsider had become part of the mainstream.

An even more recent case is that of a Greek elevator engineer, Nicholas Christophilos, who thought of a novel way to focus electron beams in particle accelerators. When, in the late 1940s, he wrote to scientists at the Brookhaven National Laboratory explaining his new idea, he was met with a long silence. The basic reason for the frosty reception was simply that scientists have long been accustomed to receiving letters from well-meaning amateurs advancing new but usually incoherent ideas. Why should this letter from Greece, filled with incomprehensible mathematics, be any different?

It turned out that this case was quite different. When the Americans finally recognized that the new idea presented by Christophilos was indeed valid, he was invited to come to the United States, where he eventually became a leader in accelerator design and in the nuclear fusion program. His original beam-focusing idea became the "strong focusing" method universally used in high energy accelerators.

Resistance to new ideas within the scientific community can arise from a number of causes. The new idea requires giving up another idea that is old and cherished, particularly if the new idea goes counter to religious teachings. The theories of Copernicus, Galileo, and Darwin are cases in point. In affairs of this type, the reasons for rejection are always emotional rather than scientific. Conflicts based on differences between science and religion continue to the present day: a large part of the population rejects the contention of biologists that a one-day old embryo is not a human being. A subset of this group continues to reject the principle of evolution. Note, however, that in science-religion disputes, it is the lay public that resists the new ideas, not the scientists.

There will be resistance if the new idea is unaccompanied by convincing empirical evidence. In 1912, Alfred Wegener, a German geologist, advanced the idea that South America originally had been a close neighbor of Africa, but had gradually drifted apart. A cursory look at the map shows that the outlines of the two continents make a close fit, and the theory seems plausible at first glance. However, the measurements of Greenland's motion that Wegener had depended on turned out to be inaccurate. The result was a great controversy during which Wegener's theory of continental drift was rejected by many geologists. Even so, Wegener's voice was not totally unheard. His effect on scientific thinking was slow but steady, and his book *Die Entstehung der Kontinente und Ozeane* ran for four editions. Today, accurate measurements of geological motion are easily made, and there is no question about the validity of Wegener's theory.

In topics such as cosmology or elementary particle theory, there is often a lack of evidence to defend particular theories. Many cosmological theories deal with events at the very beginning of time whose consequences have not yet been observed, either because the necessary instruments have not yet been invented, or because there are not enough instruments (such as large telescopes) to meet the demand. A number of elementary particle theories deal with events happening at such high energies that the accelerators required to test these theories have not yet been built. Consequently, theories by very capable scientists can exist without proof for many years, but they remain in a state of limbo until there is some pragmatic reason for believing them. These theories are not rejected outright, but neither are they accepted.

A theory that satisfies certain emotional, theoretical, and philosophical requirements may have provisional acceptance for many years, at least until experimental proof can gain it complete acceptance. Such was the case with quark theory, which built up a following for ten years until the discovery of the omega-minus baryon established the fact that the theory had the ability to predict something real.

Sometimes the new idea is couched in such unconventional and sometimes bizarre mathematical notation that conventionally trained scientists cannot or do not want to take the time to understand it. Such was the case of Nicholas Christofilos, described earlier.

The most common reason for rejection of "new" ideas is that these ideas are simply wrong (and often are not new). They defy the fundamental laws of physics and invariably come without experimental proof. The world is full of people presenting idiosyncratic ideas to the scientific community. They are certain that their ideas will revolutionize science and they cannot understand why scientists pay no attention to them. Often these ideas involve new theories of matter, or proofs that things can go faster than light. The old-fashioned perpetual motion machine is still making newspaper headlines. Indeed, there seems to be a current revival of schemes to get free power out of new kinds of electrical generators, and the traditional quack nostrums for improving the energy output of gasoline engines are in full cry. (And often these schemes are merely devices for generating money from investors who know nothing about physics or chemistry.)

Some crank ideas actually attain publication in professional scientific journals. For example, the *Bulletin of the American Physical Society* will publish all abstracts submitted to it for presentation at meetings of the Society. Almost every issue has at least one abstract whose suspect character can be recognized by its ignorance of prior knowledge, its free invention of new jargon and mathematical methods, and its inability to predict anything that can be tested by experiment. Over the years a professional scientist develops a shell of skepticism that armors him against the stream of correspondence from promotors selling their pet theories, and sensitizes him to recognize the symptoms of this pathology.

The charge that scientists have no imagination comes largely from the people who write the crank papers, making claims from discoveries that have no chance of being valid. Any scientist who refuses to waste his time poring over the blueprints for a proposed perpetual motion machine is accused of having a closed mind. Any scientist who ignores claims for ESP or UFOs or faster-than-light travel is accused of a deficiency in imagination.

I suggest that the opposite is true. It takes very little imagination to believe naively that anything is possible. Any ten-year-old child can believe this. It takes a great deal of knowledge to know what things are possible and what things are impossible.

A retrospective look at the lives of the most productive scientists indicates that each one had an interesting and hard-to-explain ability: a knack of choosing lines of research that led to important new results. These scientists avoided new ideas that were so premature or so unfocused and undisciplined that nothing could be done with them. From this kind of work no rewards flow. What the successful scientists did was to use their imaginations to decide on research topics that were on the cutting edge: problems that could be solved with the tools at hand or with tools that they could invent, and which were broad enough to represent important advances in knowledge.

Einstein demonstrated this ability for the first part of his life. His work in Brownian motion, the photoelectric effect, and special and general relativity demonstrated a wonderful ability to pick out the most fundamental and important problems of the day and to solve the problems in the simplest, most elegant fashion. The incisive paper of 1935 on the Einstein-Rosen-Podolsky paradox stimulated a discussion into the foundations of quantum mechanics which is still building momentum. Difficulties arose when he embarked on the problem of combining the four fundamental forces into one: the unified field theory. This was— and still is—one of the grandest problems in physics. It illustrated Einstein's courage in always putting himself in front of the crowd. Unfortunately, when Einstein embarked on this journey, the mathematical methods for solving the problem did not exist and he wound up spending his final decades pursuing a vain cause.

The recent case of "cold fusion" indicates that scientists do not automatically reject all unusual claims out of hand. When University of Utah chemists B. Stanley Pons and Martin Fleischmann claimed (in April 1989) that they had generated energy by the fusion of deuterium nuclei in an electrolytic solution at room temperature, the world's fusion community could have responded by saying the claim was absurd at first glance (and for good reason). Nevertheless, fusion laboratories all around the world went to the trouble of duplicating the cold fusion experiments, spending a total of $50 to $100 million in the process. None of these experiments showed a material production of energy due to the fusion of deuterium nuclei. (Some of the experiments showed tiny anomalous results, always at the edge of statistical fluctuations, making the enterprise reminiscent of parapsychology research.) The money spent on replicating the cold fusion experiment had to be taken away from funds budgeted for previously planned research. Yet it had to be done so that the world of science would not be criticized for ignoring a good idea that happened to be unorthodox.

It must be understood that for the past century scientists have been repeatedly chastised with the accusation that they are always turning down unorthodox ideas. One consequence of this criticism has been remarkable. Scientists are able to read; they are sensitive to history. Their reaction has been to become exceedingly receptive to new ideas. The pace of new discoveries in science has increased exponentially during the past century. A hundred years ago, the prevalent philosophy was: "Nothing is possible unless it is permitted." Now a 180-degree turn has occurred. The password now is: "Anything is possible unless it is forbidden." When the semantics of these statements are carefully analyzed, you realize that the two have identical meanings. The second one, however, sounds more optimistic.

Practical problems arise when available research funds must be allocated to more projects than can be funded. In the United States, most basic research money is given out by the National Science Foundation (NSF). Rolf Sinclair, head of NSF's cross-disciplinary physics programs, says that his program officers are delighted to be challenged by new ideas. When they think a proposal is receiving unfair criticism, they can step in and override the objections.[3] The NSF, in addition, has begun a program of small grants for exploratory research. Even here, however, money is not going to be given freely to ideas which have essentially no chance of working.

The question of imagination in scientists—its presence or lack thereof —is one that cannot be settled with a few anecdotes. Scientists are like any other group of people: some will be highly imaginative individuals who always look for the novel phenomenon or theory, while others will be conservative, and work within the already well-tested areas of science.

A separate question is whether scientists on the average are more or less imaginative than the general public. I think that a case can be made for the proposition that scientists are more imaginative than politicians and economists. Scientists have been in the forefront in warning about environmental dangers: the greenhouse effect, the ozone layer disappearance, radiation and chemical pollutants. The ultimate danger of unrestrained population growth was spelled out by scientists in detail twenty years ago.[4] A number of economists immediately denounced the study, insisting that continuous growth was necessary for economic strength. Economists of a conservative bent argue that we cannot afford the costs of rigorous pollution controls.

However, comparisons between disparate population groups prove very little. Some people in one group are always going to be more imaginative than some members of another group. Also, imagination is employed in different ways by dissimilar groups. In the population at large, imagination consists of inventing worlds that are the way we would like them to be. The result is literature, drama, music, folklore, myth, and religion. Scientists require a different form of imagination, one devoted to a more demanding task: discovering the world as it actually is, and finding ways to use this knowledge.

It makes no sense to complain about a lack of imagination in scientists when their failure is simply that they cannot make the world be what it is not, and they cannot make the world do what it cannot do.

NOTES

1. A. Pais, *Subtle Is the Lord* . . . (Oxford: Oxford University Press, 1982), p. 150.
2. R. W. Clark, *Einstein, the Life and Times* (New York: World Publishing Co., 1971), chap. 5.
3. E. Marshall, "Science beyond the Pale," *Science* 249 (6 July 1990): p. 14.
4. D. H. Meadows, D. L. Meadows, J. Randers, W. W. Behrens, *The Limits to Growth* (New York: Universe Books, 1972).

MYTH 9

"Scientists create theories by intuition."

Intuition: 1. the direct knowing or learning about something without the conscious use of reasoning; immediate apprehension or understanding. 2. something known or learned in this way. 3. the ability to perceive or know things without conscious reasoning.

Truth and Beauty

'Beauty is truth, truth beauty,'—that is all Ye know on earth, and all ye need to know.

John Keats, "Ode to May"

During a recent discussion relating to the psychology of scientific discovery, the astronomer Lloyd Motz, in a letter to the *New York Times,* claimed that Johann Kepler had managed to find the true laws of planetary motion even though the observational data upon which they were based did not have sufficient accuracy to support their conclusions. The data were not very important, said the writer, because the laws were inventions created through intuition. "This sort of intuition marks the true genius," he explained, "who is guided by theory rather than by 'facts' or 'data,' whose veracity is questionable. Indeed, we can interpret a 'fact' only if we have a correct theory to guide us: 'fact' thus rests on theory and not the other way around."[1]

I was startled to read this, because I thought that the habit of putting theory ahead of facts had gone out of style a few decades ago. Yet it still persists.

There is no denying that to some extent the understanding of "facts" is guided by prior theory. I have shown in another context that even a simple freshman physics experiment, intended to demonstrate Newton's second law of motion, can lead to at least two conclusions, depending on which theory you use to interpret the facts.[2] The experiment consists in pulling a mass (on a car or an air track) by a spring and measuring the acceleration. (The stretch of the spring is used to monitor the force.) If you add a second equal spring in parallel to the first, you find that the acceleration doubles. What does this experiment prove? The answer depends on how terms such as "force" and "mass" are defined, and these definitions constitute a theory by which we interpret the facts. One interpretation (the Newtonian) says: adding a second spring doubles the force, and we observe that the acceleration doubles. This experiment thus proves that the acceleration is proportional to the force. A second interpretation (the Machian) says: the force is *defined* to be proportional to the acceleration. The experiment cannot prove this fact, since it is true by definition. What the experiment does prove, instead, is that the forces applied by two springs add linearly: doubling the number of springs doubles the force.

Thus theory enables us to interpret facts. The question is: where does the theory come from?

Historically, two types of theories have been employed by scientists: (1) inductive or synthetic theories, and (2) deductive or analytical theories.

An inductive theory is established by observing regularities appearing in a large number of measurements. For example, if you do a Newton's law experiment as described earlier, and plot a curve of acceleration versus force (measured by the elongation of the spring), you might get the points to fit on a straight line. Then you could induce the theory that the acceleration is proportional to the applied force. On the other hand, if there is a lot of error in the measurements, or friction in the system, your points might not give you a very good line. However, you notice that the less friction you have, the better the curve. You could then decide to stop trying to fit the experimental points to a curve. Instead, you jump to a hypothesis: if there is no friction, then the acceleration is exactly proportional to the applied force. This is an analytic theory, created by an imaginative leap from inadequate observations. The theory is invented. If further experiments verify this theory, then you can claim your intuition worked.

In practice, the situation is often more ambiguous. When Kepler was trying to analyze the observational data accumulated by Tycho Brahe to determine the shape of the orbit of Mars, it was not just a matter of plotting the positions of Mars on a piece of paper, because the observation point (Earth) was also moving. The raw data would not have displayed an elliptical orbit. Creative intuition (i.e., use of his knowledge of geometry) led Kepler to a method of analyzing the data which portrayed the path of Mars relative to the Sun's position.[3] The true shape of the orbit still refused to emerge from the data; it was sandwiched between an eccentric circle that was too wide, and an inscribed ellipse that was too narrow. Another jump of imagination was required: by compromising between the two limiting curves, Kepler found an orbit that was an ellipse with the sun at one focal point.

Thus imagination and intuition play a role even in simple analysis of data. But do not diminish the importance of data. If later observations had not agreed with Kepler's theory, his name would not be remembered.

What do we mean by intuition? Philosophers of the past have thought of intuition as an immediate apprehension of knowledge that exists outside the mind in a Platonic world. Henri Bergson (1859-1941), a leading French philosopher and champion of irrationalism and *elan vital,* was a great believer in intuition which he described as follows: "By intuition I mean instinct that has become disinterested, self-conscious, capable of reflecting upon its object and of enlarging it indefinitely."[4] Instinct, in turn, was thought of as the opposite to intellect.

Thus, intuition was to Bergson a way of knowing without using the intellect. A definition like this is meaningless in modern usage. Terms such as "instinct" and "intuition" have little place in contemporary psychology, which does not believe in direct apprehension of the outer world. We may, if we wish, think of intuition as a metaphor describing the process of making connections between observations and ideas already registered in the memory, these connections taking place at an unconscious level. When they emerge into consciousness, the conclusions may arrive in an almost instantaneous rush. However, we do not know how much time was required by the thought processes leading to these conclusions.

To make a rude analogy with the computer model, during unconscious thinking the central processor is working, but nothing is showing up on the screen. Finally, when the process is complete, the thinker becomes aware that the end result has reached the conscious level.

In my own experience, I have often found the solution to a problem springing full-blown into my mind upon awakening in the morning. Usually this happens after a lot of mental work—both conscious and unconscious—has taken place already, but without success.

The new ideas evoked by this unconscious process seem to spring from nowhere, and we like to call them the products of free invention or creation. However, free will and free invention are never as free as we think they are. If we perform enough mental archaeology, we can often discover the basic sources of our inspirations.

To begin with, a scientist creating a new theory must start out with a purpose. This is true in any area of science. The theory must have a reason for existence. And, at least in the physical sciences, a theory must state relationships between things that actually exist. Otherwise, the theory is empty. From the outset the "freely invented" theory has some preconditions.

The wave equation of P.A.M. Dirac is often cited as an outstanding intuitive creation. This equation is analogous to Schroedinger's equation in quantum theory and describes the behavior of an electron in a given force field. Dirac's equation differs from Schroedinger's equation in that it is a combination of quantum theory and Einstein's special relativity. In some textbooks it is set down on the page as a miracle, having emerged full-blown from Dirac's intellect. Indeed, to the student the equation may seem to be a miracle because it is not derived from another equation. It is seemingly pulled out of nowhere and is written down as a general law of nature. Particular solutions for particular situations are then derived from it. Even more bewildering, the equation contains not merely algebraic quantities (such as x or y) but it is made up of blocks of characters called matrices, and the wave function is not represented by the single character *psi* (as it is in the Schroedinger equation), but is made of four components and is called a spinor.

How did Dirac come up with such a creation? Was it really a free invention? Not really. He had to begin with a motivation: a requirement for what the equation was to represent. He had to invent a wave-like equation that represented a particle whose energy and momentum were related by the equations of special relativity (rather than by Newton's laws). This procedure—a procedure guided by analogy—immediately restricted the possible equations to one very particular type. Now began the process of imagination: to invent the one equation that would fit the requirements. This invention required that Dirac draw on a knowledge of mathematics at a very sophisticated level. This process is akin to writing music or poetry. The creation cannot succeed unless there is already stored in the brain a knowledge of harmony and music theory, or a vast store of words to manipulate. You need a supply of bricks with which to erect a structure.

Having created the equation that bears his name, Dirac then proceeded to solve it—to find a psi function that represented the properties of an electron. Here comes another "miracle." Wheras the Schroedinger equation had described only the energy and momentum of the particle, the solution of the Dirac equation provided another piece of information: the spin (or angular momentum) of the particle. Putting relativity into the equation unexpectedly introduced the property of spin. Even more surprisingly, two solutions were found, with spins in opposite directions, and with opposite electric charges. Dirac's equation represented not only an electron, but a particle very similar to an electron with a positive (instead of a negative) electric charge. Did this particle actually exist? If so, where?

The answer to this question remained a mystery that persisted for four years, until (in 1932) a positively charged electron was discovered by C. D. Anderson while he was observing photographs of cloud chamber tracks caused by cosmic rays. Thus, the mysterious second solution to the Dirac equation really had a physical meaning.

The discovery of the positron assured the acceptance of the Dirac equation as a valid creation. *Without the experimental verification, the equation would have been just another mathematical trick.*

Which brings us to the role of experimental verification. Even if a theory was a totally free creation, would it have any significance without empirical validation? It could be the most beautiful theory imaginable, but if it did not allow a scientist to predict something that could be actually observed, it would just be a nice piece of mathematics or metaphysics. Today, numerous theories in cosmology and particle physics are bandied about within the community of cosmologists and particle physicists, waiting for the bits of evidence that will connect one or two of them to reality, allowing physicists to decide which among many competing theories are the correct ones.

Until the 1950s it was fashionable for scientists to claim that a theory was true because it was beautiful. Symmetry was the key to validity. Numerous symmetries had been discovered in nature, and it was believed that nature loved symmetries, so if a theory was of a symmetrical nature, this was *a priori* proof of its credence. However, in 1956, the physicists Tsung-Dao Lee (at Columbia University) and Chen Ning Yang (at the Institute for Advanced Study) discovered that it is humans who love symmetries, and that nature goes along in its own way, sometimes symmetric, sometimes not symmetric, disregarding what humans desire. One symmetry that they focused on was parity symmetry (or left-right symmetry). According to a parity symmetry, if you reflect a system in a mirror (or interchange left and right), there is no change in the behavior of the system. This leads to the law of conservation of parity, which states that in any particle reaction, the parity of the system is unchanged (parity being a number, either +1 or –1, associated with each particle). Mysteries connected with certain elementary particle reactions led Lee and Yang to question whether parity was actually conserved in reactions involving the weak nuclear interaction. Surveying the literature, they found that conservation of parity had never actually been tested in this type of reaction. Everybody believed that parity must be conserved. Because so many people had said that nature loves symmetry, we all came to believe it.

Very quickly an experiment on beta decay, performed by Madam C.S. Wu and others at the National Bureau of Standards, demonstrated that parity was not conserved in this reaction, and that, as a result, it was possible to distinguish between the left hand and the right hand (and between clockwise and counterclockwise) in an absolute manner. Nature was not symmetrical with respect to left and right in reactions mediated by the weak interaction.

This result knocked the idea of truth and beauty into a cocked hat. It showed that any hypothesis created through the imagination must withstand rigorous testing, and that this testing must be done separately for each of the four fundamental interactions.

Albert Einstein understood well the relationship between theory and experiment. At the Herbert Spencer lecture, delivered at Oxford University on June 10, 1933, he said: "Pure logical thinking cannot yield us any knowledge of the empirical world; all knowledge of reality starts from experience and ends in it. Propositions arrived at by purely logical means are completely empty as regards reality. . . . The structure of [a theoretical system of physics] is the work of reason; the empirical contents and their mutual relations must find their representation in the conclusions of the theory. In the possibility of such a representation lie the sole value and justification of the whole system, and especially of the concepts and fundamental principles which underlie it. Apart from that, these latter are free inventions of the human intellect, which cannot be justified either by the nature of that intellect or in any other fashion *a priori.*"5

It would be easy to take an out–of–context quote from this paragraph and claim Einstein said, "physical theories are free inventions of the human intellect," which is far from what he actually said. The meaning of the entire quotation is that theories result from a complex chain of events; the free invention of fundamental concepts and principles are only the first step in this chain. The use of reason to hypothesize theoretical systems is a later step, after which the theory is connected to reality by the deduction of physical consequences. A proper theory must say that something observable is going to happen. We then look to see if those deduced consequences actually refer to events occurring in nature. If observations of nature agree with the predictions of the theory, then the theory is verified. To state it more concisely: the theorist is free to make up all the theories he can imagine. It is up to the experimenter to decide which of these many theories describe the real universe. (A full description of theory-making would include the feedback cycles taking place when the differences between the predictions and the observations force the scientist to modify and correct the theory.)

A sense of beauty and proportion may assist in the initial creation of a theory, but no matter how beautiful a theory is, if it doesn't work, it is not a good theory. If intuition were as all-important as is claimed, and if data were as unimportant, we might wonder why scientists have spent so many billions of dollars over the past several decades to build particle accelerators, particle detectors, telescopes of all kinds, and the rest of the paraphenalia of experimental science. The answer is simple: theory without empirical evidence is akin to art, poetry, literature, and music. It may have beauty and symmetry, but you don't know that it has truth unless you compare what is inside your mind with that which is outside your mind.

The career of P.A.M. Dirac demonstrates how overemphasis on beauty as a guidepost to scientific discovery can lead a scientist astray. We have seen that Dirac's invention of the wave equation that bears his name was a prodigious exercise in creativity. Its success reinforced Dirac's belief that the road to scientific progress was through mathematical beauty. Unfortunately, Dirac's quest for beauty did not serve him well following that initial achievement. Although he continued publishing until 1984, Dirac's work after 1935 met with limited success, and gradually he began to be thought of as out of the mainstream.[6]

The story of Dirac and his quest for beauty is another example of the failure of idealism. Although Dirac's equation is indeed beautiful in the eyes of a physicist, it is a successful equation not because of its beauty, but because it is a realistic equation.

NOTES

1. L. Motz, Letters to the Editor, *New York Times* (13 Feb 1990).
2. M. A. Rothman, *Discovering the Natural Laws* (Garden City: Doubleday, 1972; New York: Dover, 1989), chap. 2.
3. E. M. Rogers, *Physics for the Inquiring Mind* (Princeton: Princeton University Press, 1960), p. 265.
4. B. Russell, *A History of Western Philosophy* (New York: Simon & Schuster, 1945), p. 793.
5. A. Einstein, "On the Method of Theoretical Physics," in *Ideas and Opinions* (New York: Dell Publishing Co., 1954).
6. H. Kragh, *Dirac: A Scientific Biography* (New York: Cambridge University Press, 1990).

MYTH 10

"All theories are equal."

A Democracy of Science

Henry M. Morris, Director of the Institute for Creation Research in San Diego, a leading exponent of "scientific creationism," begins a book on the "scientific basis" for creationism with the statement that evolution and creationism are the only two models capable of explaining human existence.[1] He then claims that it is impossible to prove scientifically which model is correct, since creation is not taking place now, and so is not subject to experimental observation, while evolution takes place so slowly that it could not be observed either. Since it is impossible to prove which model is correct, it follows that both models are equal, and thus merit equal treatment in the schools.

Dr. Morris also claims the following: "If the neo-Darwinian theory is axiomatic, it is not valid for creationists to demand proof of the axioms, and it is not valid for evolutionists to dismiss special creation as unproved as long as it is stated as an axiom."[2] In other words, two axiomatic theories must be considered on an equal footing (even though they imply diametrically opposite conclusions), since the underlying axioms are free inventions of the human mind and thus are not susceptible to proof.

This example demonstrates an insidious consequence of the myth discussed in the previous chapter: *Scientists create theories by intuition.*

If theories are nothing more than deductions from axioms that are intuitively true, then all theories are equally valid.

While such a philosophy has some believability when it is applied to pure mathematics, it is a parody of the truth when applied to the natural sciences. In mathematics, Euclidean geometry and any of the competing non-Euclidean geometries have equal validity, even though they are based on different sets of axioms. But when we ask which of these geometries applies to plane figures on a plane surface, then only the Euclidean version passes the empirical tests (such as agreeing with the Pythagorean theorem). Proceeding further, when we ask which geometry applies to the real universe, we find that any scientific theory which is to be accurate over very large distances must use one of the non-Euclidean geometries.

In any of the natural sciences, a theory must not only be logically consistent, it must also be subject to reality testing. That is, the theory must say something about the real world that can be observed, and the predictions of the theory must agree with the observations. This means that a scientific theory must pass three independent tests before it can be accepted as a correct theory:

1. The theory must allow for the application of Karl Popper's test of falsifiability. That is, the theory must predict that something real can be observed, so that if this thing or event is *not* observed the theory is proven to be false. In other words, the theory must, in principle, be falsifiable.

2. For the theory to be correct, there must be no empirical evidence that disproves or falsifies the theory.

3. Finally, there must be some empirical evidence that supports the theory and lends credence to its validity.

The first test is often cited as the only thing necessary to make a good theory. In this simple view, falsifiability is the only requirement for raising a theory to an intellectual level worthy of entrance into the academic curriculum. This attitude is grossly inadequate and incorrect. Falsifiability only proves that a theory is an *empirical theory*—a theory about the real world that is capable of verification. By itself, falsifiability does not guarantee that a theory is valid. Newton's laws of motion, for example, make up a falsifiable theory, but in the domains of high velocities or very small masses, it is not a valid theory.

What happens to creation theory when we apply the test of falsifiability to it? First, we must find a reasonable definition of creationism. According to Morris, the creation model starts with a period of *special* creation in the beginning, during which the basic systems of nature were brought into existence in completed, functioning forms. Since "natural" processes do not accomplish such things at present, these creative processes must have been "supernatural" processes requiring an omnipotent, transcendent *Creator* for their implementation.[3] In addition, the creation took place very suddenly, via a great catastrophe. As Morris says, "If the Creation-Flood Model is valid, then there is no real reason to think the earth is much older than mankind and the beginning of human history."[4] Let us be generous and say that the creation took place no more than 10,000 years ago. (If we take the Biblical creation to be literally true, then the creation would be closer to 6,000 years ago.)

How do we falsify a theory of this kind? First, we know that dinosaur fossils have been found within the earth's surface. We can determine, using radioisotope dating, that these fossils originated millions of years ago, far more than the 10,000 years allowed by creation theory. The existence of these fossils thus falsifies the theory. But no, say the creationists. The dinosaur fossils were put in place less than 10,000 years ago by the omnipotent Creator, giving the illusion that the earth is older. The dating systems used by evolutionists are no good for long time spans because rates of radioactive decay were suddenly changed a few thousand years ago by this same Creator, with the result that the fossils seem to be older than they actually are. Creation theory is such that any time a practical objection to creationism is raised, the omnipotent Creator is brought in to explain away the objections. The *deus ex machina* produces a *reductio ad absurdum*!

With this kind of logic you can prove anything you want. Creationism is a theory that cannot be falsified, because it permits the invention of *ad hoc* supernatural explanations for any observable fact. The theory is designed so that no matter what kind of evidence is produced to show that the earth is more than 10,000 years old, the creationist can say this evidence came into being by the will of the Creator. It is thus impossible for the evidence to show that creation theory is false.

As I showed in chapter 4, this kind of logic is circular logic. It assumes what it is trying to prove. The basic premise of creationism (the omnipotent Creator) is designed so that any opposing evidence is arbitrarily, without any rule of law, contradicted by the theory.

Since creationism cannot be falsified, it is not an empirical theory. It is not a theory that applies to real things in the real universe. Since creationism is not an empirical theory, it cannot be a scientific theory, regardless of the attempts of its adherents to give it the title "Creation Science." It is simply fantasy, as are all non-empirical theories that pretend to describe the real world.

That is the significance of the first test: a theory must be falsifiable in order to be empirical, and it must be empirical if it is to be verified by real observation. Furthermore, if the theory does not pass the first test, then there is no point in applying the next two tests, because they cannot give meaningful results. Evidence is meaningless if it can be explained away by ad hoc supernatural explanations.

How does the theory of evolution hold up when the three tests are applied? Evolution does allow one to make specific predictions about things happening in the real world. For example, if evolution is a correct theory, human remains should never be found together with dinosaur fossils, because the two animals lived at very different times in history. This is a simple prediction, but that does not make it trivial. Even before we attempt to verify the prediction, the very fact that the prediction can be made, and that the appropriate observations are possible, assures us that evolution is an empirical theory.

But is it a valid theory? That is a separate question. To answer this question we must look at what we find in nature. Indeed, we do not find human remains in fossil beds together with dinosaur skeletons, so the theory is not only falsifiable, but it is not falsified. It has passed the test of rule two. (Of course, to test the theory thoroughly, we should make all the predictions we can think of and then see if these predictions are true or false. If none of the predictions are found to be false, the theory is unfalsified and is provisionally acceptable.)

So far so good. However, before we can totally accept a theory, we need more than a lack of falsification. Validation requires verification. If evolution is true, it should be able to predict something that is going to happen or has actually happened. We call it retrodiction when a theory predicts something that already happened. The most definitive test is that we should be able to observe some organisms in the act of evolving.

In the case of microorganisms, we see evolution taking place all the time. We see bacteria developing into strains resistant to specific antibiotics when they grow in a medium containing those antibiotics. A more persuasive test would be an observation of evolution taking place in higher organisms. Recently, such a test was reported by David A. Reznick, Heather Bryga, and John A. Endler of the University of California.[5] Their experiment verified a mathematical theory that predicted evolutionary change in animals determined by the number and type of predators in their environment. The experimenters transferred a number of guppies from a river that contained fish whose habit was to eat large mature adult guppies, to a river whose predators preferred the small newly-born guppies. After 11 years and 30 to 60 generations, the investigators found that the guppies had changed in accordance with the theory. The fish transferred to the river without the adult-eating fish had evolved so that they matured later and had fewer and larger offspring. The value of this experiment was not only in the fact that an evolution had taken place, but that the nature and amount of the change agreed with the mathematical predictions that had been made ahead of time.

Evolution thus passes all three requirements of a scientific theory, while creationism does not even pass the first test. Yet in various parts of the country, vociferous groups are attempting to install creationism as part of the science curricula in public schools under the pretext that both theories are equal. While they have consistently been defeated in the courts, the chilling effect that remains is seen in the fact that evolution is inadequately treated in many textbooks.

The reason a portion of the public supports creationism is because it believes the myth that "theories are creations of the mind," and that therefore, there is no way of proving one theory is better than another. It is part of the democratic spirit. A related anti-elitist myth informs us that nobody is smarter than anybody else. Therefore, nobody can create a theory that is better than any other. Egalitarianism, carried to extremes, breeds an atmosphere inimical to clear thought.

There is also a hint of philosophical idealism in the myth. If theories are nothing more than creations of the mind, then empirical evidence is of little importance. While creationists give lip-service to observations and data, the way they interpret what they see is always in favor of the theory they carry in their minds. Evolutionists, on the other hand, are realists. Their theory was invented to fit observations of the real world; it does not call on forces from outside the universe to explain the nature of the world.

Evidence of the split between idealists and realists can also be seen in the ongoing struggle between the two sides in the abortion debate. The basic premises of the anti-abortion side are that "life is sacred" and that "life begins at the moment of conception." These are purely idealist positions. The first originates in religion and the second is an invention of the "pro-life" faction with no relation to reality. For nothing *begins* at the moment of conception. What takes place is the fertilization of an egg. Both egg and sperm were alive before conception; life did not begin at that instant. Conception is nothing more than the joining of two sets of chromosomes, or DNA molecules. You may, if you want, define the resulting embryo to be a real, individual human being whose life began at the moment of conception (even if it has only one or two cells). You have the right to make that definition. But do not try to claim, as many do, that science "proves" the embryo to be a live human being. The moment of time when the embryo becomes a particular person is entirely a matter of definition and is not subject to scientific proof.

For example, we may define the borderline between embryo and human being to be the time when the synapses of the nervous system are completely formed, so that the brain is capable of real thinking. This does not happen until the third trimester. Pinpointing the time when this happens is a matter susceptible to scientific investigation, but the judgment that this instant is the beginning of a human being is simply a matter of personal choice.

We see in this conflict a classical conflict between opposing theories. The first theory says a human being starts to exist at conception. The second theory says a human being starts to exist when the brain is capable of sustaining thought. These are not empirical theories, because they depend on the definition of "human being." Are both definitions equally good? I think not, because one is an idealistic theory, depending on a theological definition of "human being." (Does the soul enter the fertilized egg at conception?) The other is more realistic, giving a pragmatic test for determining when the "human being" actually exists.

These examples demonstrate that the question of idealism versus realism is not merely an abstract debate between ivory-tower philosophers. It is a political battle fraught with enormous consequences. Much symbolic blood, and some real blood, is spilled because of basic differences between idealists and realists. To idealists, thoughts are more important than material objects. Consequently symbolism plays an exaggerated role in their thinking. To the idealist, the symbol is the same as the thing it represents. Therefore, burning a flag is like burning the country and the constitution. To the realist, a flag is a piece of cloth, and the fact that it has great meaning and emotional reverberations is recognized and understood, but is not allowed to overwhelm the debate. The idealist is more outraged by a burning flag than by the sight of homeless people living on the streets. The realist is more interested in correcting real defects than in punishing symbolic gestures.

NOTES

1. H. M. Morris, *The Scientific Case for Creation* (San Diego: Creation-Life Publishers, Inc., 1977), p. 8.

2. C. L. Harris, "An Axiomatic Interpretation of the Neo-Darwinian Theory of Evolution," *Perspectives in Biology and Medicine* (Winter 1975): p. 179.

3. Morris, *The Scientific Case for Creation,* p. 3.

4. Ibid., p. 43.

5. G. Kolata, "Living and Dying in the Wild, Guppies Back Evolution Idea," *New York Times* (26 July 1990): p. A1.

MYTH 11

"All scientists are objective."

Stereotype: an unvarying form or pattern; specifically, a fixed or conventional notion or conception, as of a person, group, idea, etc., held by a number of people, and allowing for no individuality, critical judgement, etc.

Stereotypes

Modern entertainment—science fiction movies in particular—has given rise to a number of stereotypes about scientists that swirl about in the public mind:

The scientist as selfless, tireless, and noble. We have not seen this image in films recently, but movies of the 1930s and 1940s abounded in stories about the nobility of scientists, a stereotype which I do not mind. Paul Muni played Louis Pasteur, the discoverer of bacteria as the source of disease. (*The Story of Louis Pasteur,* 1936.) Edward G. Robinson abandoned his gangster roles to play Paul Ehrlich, developer of salvarsan, the "magic bullet" which killed the syphilis spirochete and became the first successful chemotherapy agent. (*Dr. Ehrlich's Magic Bullet,* 1940.) Greer Garson represented Marie Curie, the discoverer of radium and polonium, a truly noble woman who refused to patent her discoveries because she thought they belonged to all mankind, an attitude as rare today as it is remarkable. (*Madame Curie,* 1943.)

In all of these stories, the scientist was depicted as hard-working, single-minded to the point of obsession, struggling to overcome material obstacles and human stupidity, finally bringing forth a discovery of immense value to the world. The image is one that could be useful at the present time, when it is difficult to interest most young people in the sciences—indeed, when it is difficult for most students to learn what it is that scientists really do, unless they have a parent in the field.

More often, the popular stereotype of the scientist is the image whose roots can be traced to the Faust legend: the scholar who sells his soul to the devil for eternal youth, for power, or for other personal gain. From this arises the "mad scientist," renowned in science fiction, typified in film by Dr. Frankenstein, Dr. Jekyll, Dr. No, Dr. Strangelove, and Rotwang, creator of the evil robot in *Metropolis*. The universality of the mad scientist in science fiction is somewhat baffling, since history has not yet seen a paranoid scientist actually using his abilities to attempt a world conquest. Soldiers and politicians, yes, but not scientists.

It is true that many scientists gain a livelihood designing weapons, and some seem to enjoy it more than others. It was rumored that the character of Dr. Strangelove was modeled after a well-known physicist whose chief claim to fame was his championing of the hydrogen bomb and the "star wars" strategic defense initiative. But his motives were defensive, rather than offensive.

Since there is no factual basis for the mad scientist, we must consider it a prejudicial fantasy similar to those directed at ethnic minorities. It originates from the perceived difference between the scientist and the rest of society. The scientist is "brilliant." He knows secrets of nature that are hidden from the masses. He is probably an atheist and therefore is not to be trusted. He unleashes destructive forces on the world: radioactivity, electromagnetic waves, chemical pollution, genetic engineering, etc.

Never mind that scientists are the ones who originally disseminated warnings about environmental problems. The essence of stereotypic thinking is that differences within a group are ignored: all members of the group are alike.

The stereotype of the scholar as evil, if not mad, is far older than the cinema. It begins in the Garden of Eden, with dire results emanating from the knowledge of life taught by the serpent. From Ecclesiastes we learn that "He that increaseth knowledge increaseth sorrow." Knowledge itself is evil, for it goes against the myth that "there are things on earth man was not meant to know." To "know" itself has sexual and, therefore, forbidden connotations. The medical doctor has intimate knowledge of bodily parts and functions considered "dirty" by others. The biochemist, on the verge of knowing what life is, is making supernatural forces superfluous. The physicist and the cosmologist seek to know the origin of the universe, making the Bible superfluous. (Is it possible that some of the resistance to the superconducting supercollider project originates in a reluctance by non-scientists to probe too deeply into the fundamental structure of matter?)

The opposite stereotype is that of the scientist as "nerd." Far from desiring power, the nerd is meek and mild, with a low emotional level and a minimum of assertiveness. He wants nothing but to be left alone with his computer. Occasionally he rebels and has an adventure, but in the main he shuns conflict and withdraws to his books. He wears horn-rimmed glasses, he is slightly stoop-shouldered, and he is not very interested in girls. He haunts science fiction clubs and actually reads books.

Nobody knows how many bright students are lost to science by the dissemination of the nerd stereotype throughout our culture. It takes a strong will to overcome the prejudices of the herd. But enough students do resist to fill the annual science competitions.

You may wonder how it is that the same group of people can inspire three stereotypes as opposite in character as the above. The reason is simple: the people who engage in stereotypical thinking rarely know the people they are making judgements about. This kind of thinking makes no allowances for individual differences between the members of the group being stereotyped. Stereotyped thinking about the black population "knows" that blacks do jazz and sports. It has no room for the idea that some blacks are symphony musicians and others physicists.

In the same way, a physicist who likes to play baseball, and a physicist who is so religious that he will not use a telephone on Saturday, are individuals who fit no stereotypes. Neither of these is a mad scientist or a nerd.

A fourth stereotype running through modern mythology is the idea that a scientist at work is totally objective. He is a kind of walking computer who does not let feelings get in the way of his ideas or observations. This myth is different from the two mentioned above. It is not supposed to be derogatory. It is supposed to be a compliment.

Yet it is, regretfully, not always true.

Because it is the main topic of this chapter, let us discuss this idea in more detail.

The Battle Against Delusion

The stereotype of the scientist as cold-blooded and totally objective evokes the image of a dour scholar unemotionally going about his business, compiling facts and drawing indisputable conclusions. Some fictional detectives summon forth a similar fantasy: Sherlock Holmes smoking his pipe, rationally deducing a criminal's identity; Peter Falk as Columbo intoning, "Just looking for the facts, ma'am." However, a serious question arises: which of many facts does the scientist choose to see?

The simplest organization of facts into patterns requires choices. A biologist arranging living creatures into a taxonomic system must first choose his criteria: is the classification going to be according to the number or type of limbs, according to body temperature, or according to the presence or absence of a spine? Making the choice implies a theory, even if only a rudimentary one. We must then ask whether the theory is a useful one, and if so, what is its purpose? If the classification scheme is intended to support the theory of evolution, then an observer in favor of evolution might tend to emphasize one set of facts, while an opponent of evolution would tend to favor another set of facts. The evolutionist, for example, looks at fossils of primitive horses and sees that they form a rather complete series, demonstrating the process of evolution in action. The creationist, on the other hand, looks at the same fossils and sees only the gaps in the chain, the missing links.

The myth of objectivity encourages people to imagine that scientists are like computers, with the ability to put aside their desires and feelings. The fact is that scientists are human beings who share the foibles and failings of all human beings. When these human scientists start to generate theories, they begin with fantasies that only gradually become refined into rigorous principles. The scientist always begins by wanting to explain something. Whether to explain the whole universe or to explain one small observation, his fundamental desire is to explain. Next comes his desire to prove the explanatory theory, to show that it is a correct theory.

Here is where the danger begins. If a scientist becomes very excited about a new theory, and if he tries very hard to prove his theory, he may run into the danger of losing objectivity, of becoming so emotionally involved with his research that he starts picking and choosing among the experimental results from which he obtains his data. Consequently, he may be tempted to throw away displeasing data, and to keep only the data that proves his theory. This can be a very subtle process and it is not always dishonest. Sometimes an experiment does give meaningless data. Something goes wrong with the equipment, or noise gets into the electronics. Therefore, there can be a legitimate reason for discarding numbers that don't fit the theoretical curve well enough. A scientist aware of such dangers spends some time trying to disprove his theory. If he is unable to disprove it, then his proofs are even more convincing.

Analyzing sources of error in an experiment is perhaps the most difficult part of laboratory work. Often the errors are underestimated, so that the apparatus is believed to be more precise than it actually is. A well-known example is the fact that the first measurements of the speed of light reported velocities that were uniformly greater than the results of more recent measurements. Looking at the data, one might be tempted to conclude that the speed of light has gradually been decreasing during the past century. Such a turn of events would wreak havoc with all the most fundamental concepts of physics, with our vision of the universe and its structure. Fortunately for the constancy of the speed of light and the theories that depend on it, it has been shown that the error bars put on the early measurements by optimistic experimenters were smaller than the true margin of error. Measuring methods have gradually improved in their accuracy, and now they are converging on the correct value for the speed of light. The early measurements were simply less correct than the scientists had thought.

This, in itself, was a perfectly innocent error. There was no reason for any scientist to prefer one light-speed more than another. There was no theory requiring the speed of light to be any particular number, therefore, no theory to fall in love with and no subjective biases to interfere with the measurements.

In numerous other instances, however, wishful thinking overcame objectivity. One of the most flagrant examples of shared delusion was the case of three astronomers—Giovanni Schiaparelli, Nicolas Flammarion, and Percival Lowell—who thought they saw canals on Mars for so many years. They—in one way or another—arrived at the belief that the surface of Mars was covered with canals, and so duped themselves into believing they saw these canals through their telescopes. Photographs did not show the canals, and most professional astronomers did not believe in their existence. Yet the canals and the advanced civilizations that must have built them remained the meat and potatoes of science fiction for many years, at least until space vehicles photographed Mars from close orbit in 1975.

The most notorious example of experimenter error in recent years is the case of cold fusion.[1] This was the experiment which purported to demonstrate the controlled fusion of deuterium nuclei within palladium electrodes during the electrolysis of heavy water at room temperature. The errors were of several kinds.

First—and perhaps most important—the two chemists making the claim (B. Stanley Pons and Martin Fleischmann, at the University of Utah) reported their work to the press on March 23, 1989, before it had been published in a peer-reviewed technical journal. Their stated reason was that rumors of their results had already been leaked and they wanted to claim priority for patent purposes. The result of this unprofessional procedure was that nobody knew exactly how the experiment had been performed, so those attempting to replicate it had great difficulty knowing what the precise conditions were. After all, the purpose of publication is to permit replication, and replication is what keeps scientists honest.

Second, the chemists claimed that their experiment showed a measurable emission of energy in the form of heat. This is what made the affair so important. Nuclear fusion is a possible source of energy with an essentially inexhaustible fuel supply. Since the early 1950s, a vast and expensive research project has been carried out all over the world to generate power by heating deuterium and tritium gases to extremely high temperatures—a process known as thermonuclear fusion. If fusion could be made to take place at room temperature, knowledge of the method would be of indescribable value.

The theory of the Utah group was that fusion of deuterium nuclei could take place if the deuterium gas was dissolved in palladium, a metal long known to dissolve hydrogen and its isotopes in large quantities. However, attempts to duplicate the work of the Utah group in fusion laboratories around the world failed to generate measurable amounts of energy.

It is easy to see from published reports that errors in experimental techniques were being made by both the Utah group and others attempting to repeat the work. Precise measurements of heat under the dynamic conditions of the experiment are very difficult to perform. A detailed analysis of the Pons and Fleischmann procedure demonstrates how errors in analyzing the data could have led the Utah experimenters into believing that more energy was coming out of the electrolysis chamber than was going in.[2]

Third, it is well known that neutrons must be emitted during the fusion of deuterium nuclei, and it is a simple matter to calculate how many neutrons should be emitted each second for each watt of generated power. Such large numbers of neutrons were never seen. One person did claim detection of copious neutron emission, but it turned out that his neutron detector was malfunctioning, and the claim was withdrawn. A report of this nature is evidence of gross negligence. Any person claiming to be an expert in radiation detection knows that the first thing you do before making a measurement is to test your detector with a calibrated radioactive source. (In the case of neutrons, one uses a polonium-beryllium source or its equivalent.) Had this elementary precaution been taken, it would have been impossible to rush to publication with counts from a faulty detector.

Many experiments claimed occasional generation of heat, but no neutrons. Others claimed generation of small bursts of neutrons, but not a lot of heat. The conflicting and sporadic results is evidence that errors were being made. A possible source of spurious neutron counts was identified by one team from the Sandia National Laboratories who showed that when counting low-level neutron fluxes, a single neutron counter can create the illusion of neutron counts because of high-voltage breakdown. Essentially what you see is electrical noise. Therefore, one must use arrays of several counters for reliable detection of low-intensity neutron emission.[3]

The main factor that contributed to the publication of erroneous results was the strongly-held belief that important scientific discoveries were being made. Clearly, cold fusion might have been an extraordinarily important development. With the end of oil production looming during the next century, the discovery of a new and cheap source of energy would be of incredible monetary value. The belief that a bottomless energy supply had been found sent a surge of wishful thinking and recklessness through a large number of minds. Visions of valuable patents dazzled the minds of many, distorting their judgments. It was the responsibility of the scientists involved to slow down, to take a second look, to make sure that they had analyzed all possible sources of error, and then to publish the results in the most prestigious peer-reviewed journals. Instead, the experimenters went flying to the press with absurd claims based on hasty measurements. The atmosphere was one of uncontrolled hysteria.

Most interesting, in view of the presumed objectivity of scientists, is the manner in which experimenters and observers alike lined up on both sides of the question, as though they were choosing teams at a soccer match. Most of the cold fusion supporters were chemists, while most of the skeptics were physicists. This is not simply a matter of group loyalty. It has to do with the differing ways in which chemists and physicists think about the world. Many chemists (but not all) tend to look at things from above, from a macroscopic, top-down point of view. They are accustomed to thinking in terms of chemical reactions between masses of substances. Most physicists, on the other hand, habitually think in terms of reactions between individual atoms, nuclei, or elementary particles. They can tell you immediately what kind of nuclear reactions are required for a fusion reaction to take place, and they were highly skeptical about claims of fusion reactions *without* the emission of enormous numbers of neutrons. (The initial newspaper accounts did not even mention neutrons, and when they finally did, they did not specify what method Pons and Fleischmann used to measure them.)

In spite of their skepticism, the physicists did not immediately denounce the cold fusion experiments. They took the time to repeat the experiment with their own apparatus and found that nothing important happened. The cold fusion enthusiasts maintained to the bitter end that something new and strange was going on. So the dispute continues. But one thing we know for sure. Cold fusion is not going to solve the energy problem. The process is not going to be a source of power. People here and there are going to claim generation of small amounts of neutrons or production of small amounts of tritium within the palladium. There will be (and already are) claims of tritium contamination in the palladium. But all this is only of academic interest.

(At the moment of writing, the newspapers report that Stanley Pons has disappeared, and that he will speak to no one—except through his lawyer—until his patent claims have been settled. This behavior verifies the hypothesis that the scientific pathology evident in this case was caused by anxiety over being the first to file patent claims. It is in ironic contrast to the attitude of Marie Curie, who refused to claim patents on any of her discoveries, and who received two Nobel prizes as a reward.)

The effects of experimenter bias is a subject that has been studied for many years by psychologists and biostatisticians alike. The scientists who first gave thought to designing experiments so that the experimenter could not influence the data were medical researchers engaged in testing new drugs. A century and a half ago these scientists realized that when a physician administered an experimental drug to a number of patients, his expectations had a noticeable effect on the apparent effectiveness of the drug. There were two major reasons for this effect: (1) If the administering doctor had great faith in the drug, then his behavior would engender a sense of optimism in the patients, causing them to feel better regardless of the efficacy of the medication. Or, (2) the patients, noting that the doctor expected good results, would claim they felt better in order to make the doctor feel happy. Both of these effects are nothing more than variations on the stage magician's traditional use of suggestion.

In response to this problem, members of the Vienna Medical Society invented the double-blind method of research, putting it into use as early as 1844.[4] In the double-blind method, the patients are divided into two groups. One group gets the new medication, while the second group (the control group) receives a placebo—an inert substance that produces no effect other than a psychological one. If the control group is observed to receive as much benefit as the group taking the actual drug, then the drug is no better than the placebo—that is, no good at all (unless you are satisfied with a placebo effect). The essence of the double-blind method is to avoid the effect of experimenter bias by ensuring that the doctor does not know which medicine the patient is given, and that the patient does not know which medicine he is receiving. This result is obtained by having the medication (identified only with a code) packaged anonymously by a third party who is isolated from both the administrator of the drug and the recipients.

Numerous studies, involving thousands of subjects, have demonstrated that the double-blind method really is effective in reducing experimenter bias. As a result, this technique has become the method of choice in all kinds of research involving the responses of groups of people to various kinds of stimuli. Psychology research has benefited from it, and in parapsychology research the use of the double-blind method is mandatory. In using the double-blind method in parapsychology work, the person recording the data must not know what kind of signal the subject is receiving, and must not know whether or not the subject belongs to a control group.

In spite of modern knowledge of psychology, spread to the public through popular magazines, trickery in the employment of suggestion still prevails in many quarters, particularly where there is money to be made. A recent fad has been the selling of audio tapes containing "subliminal messages" in order to induce therapeutic effects such as an increase of self-esteem, or the improvement of memory. (Subliminal messages are messages so brief in duration that they do not reach consciousness, but are supposed to produce changes in behavior.) The naive use of such tapes might be called an uncontrolled experiment.

However, a variation of the double-blind method was put to good use in one experiment performed by Anthony Greenwald, a psychologist at the University of Washington.[5] In his study, a number of volunteers listened to a tape that was supposed to boost self-esteem. Another set of volunteers was given a tape that was advertised as improving one's memory. What they did not know was half the volunteers received tapes with the labels reversed. When they were asked if they believed their memory or self-esteem had improved since using the tape, about 50 percent said that they had indeed experienced an improvement. This improvement, however, was a result of the label on the casette, rather than the message in the tape. When objective measures of self-esteem or memory were taken, there was essentially no change.

The mechanism of autosuggestion explains the positive results obtained by dubious experiments when the scientists performing the experiments strongly believe that the experiments are going to prove their fondly-held theories. This is not to disparage the importance of enthusiasm. Enthusiastic belief in a theory is the hallmark of the dedicated scientist, and is necessary for the scientist to function. However, there are times when unbridled enthusiasm unhinges the judgement of the experimenter and causes him to accept results that a more cautious worker might discard. This is especially risky when the experiment gives results on the edge of what might be expected by chance.

Numerous cases of self-delusion by professional scientists have been documented elsewhere.[6] Some call this phenomenon "pathological science" and others, not wishing to invoke psychopathology, use the term "wishful science." While these pathological events do not make up a large fraction of total scientific discovery, they are important enough, and occasionally bizarre enough, so that attention must be paid to them.

Some psychologists have studied the particular mechanisms by which legitimate scientists make theoretical or experimental errors, and they agree that too many times the potential for error is fueled by overenthusiasm for a particular theory. Theodore Barber has listed ten pitfalls which lie in wait for the unwary experimenter.[7] While he aims his analysis mainly at those working with human subjects, his comments can be applied to any kind of research which relies on the amassing of experimental data, especially of a statistical type. In his discussion, he distinguishes between the investigator and the experimenter. The investigator is the person who originates, designs, conducts, and interprets the experiment. The experimenter is the one who personally does the tests, records the data, and performs the data analysis. Of course, in a small group the same person may play both roles. The ten pitfalls analyzed by Barber are the following:

1. *Investigator Paradigm Effect.* A paradigm is a set of shared beliefs among scientists, the conceptual framework of a theory. In his famous book, *The Structure of Scientific Revolution,* Thomas Kuhn analyzed the shift of paradigms that takes place in science from time to time.[8] In particular, the change from the earth-centered to the sun-centered solar system was one of the major shifts in the history of science. More recently, the shift from Newtonian mechanics to quantum and relativistic mechanics has changed the paradigms of physics from classical to modern modes. Since a paradigm is so pervasive in the scientific culture, the mindset it produces in the researcher is often unconscious and so is difficult to compensate for.

In experimental research, the prevailing paradigm can influence the kind of questions that will be asked, which, of course, will determine the kind of answers that are received. Even in physics a paradigm effect may be important. Before the discovery of the positron, physicists occasionally saw tracks in their cloud chambers that looked as though they were made by electrons but which curved in a direction opposite to electrons in a magnetic field. But nobody could imagine that these tracks were actually made by positively charged electrons. Everybody knew that electrons were always negative. Laborious calculations were made to show that multiple collisions with the atoms in the cloud chamber gas might possibly produce an anomalous curvature. After Dirac made the possibility of a positively charged electron conceivable, the simplest answer turned out to be the correct one: the strange particles in the cloud chamber were positrons.

A similar paradigm effect was responsible for the fact that prior to 1962, no chemist had ever looked for compounds of krypton and xenon because the textbooks of the period unanimously said that the noble gases never entered into chemical combination. To their embarrassment, it was found that a number of such compounds could be made in a fairly simple manner. It is difficult to overcome what you learn in freshman chemistry.

2. *Investigator Experimental Design Effect.* In psychology, it is found that simple experimental designs give simple results, while complex designs give more complex results that are sometimes at odds with the simple ones. Early experiments on hypnosis tested the suggestibility of subjects under a hypnotic trance compared with the suggestibility of subjects in an awakened state. Hypnosis indeed was found to increase the suggestibility. Later experiments added a third condition: the subjects were urged to try to perform to the best of their abilities and to imagine vividly the things that were suggested. It was found that the "task motivational condition" raised the suggestibility of the subjects as much as the hypnotic state, showing that there was nothing particularly unusual about hypnosis.

3. *Investigator Loose Procedure Effect.* It is well known that in studies based on polls or questionnaires, the answers received depend on how the questions are worded and on how the subjects are chosen. An infamous poll that predicted the election of Thomas Dewey over Harry Truman in 1948 was skewed in its statistics because it depended on telephone interviews for its data, and in those days more Republicans tended to have telephones than Democrats because Republicans tended to be wealthier than Democrats. The pollsters, as well as some newspapers that had already printed their headlines the previous night,

were greatly chagrined at the election results.

It is important for experimental procedures to be specified carefully in advance and followed closely. In particular, when obtaining data that is statistical in nature, it is important to make a hypothesis and decide what you are looking for prior to taking the data. Furthermore, the investigator must decide ahead of time how much data is to be taken, when the data-taking starts, and when it stops. Obviously, if a run of good results is encountered and the investigator then decides it is a good time to stop, the statistics are going to be biased in favor of the hypothesis. This Loose Procedure Effect is especially important when the investigator is looking at very small effects in the midst of large amounts of noise and so is a prime factor in parapsychology research.

In addition, the investigator must decide ahead of time on the wording to be used if the study involves asking questions. The manner of questioning can have a great effect on the answers received, as every fortune-teller knows only too well.

4. *Investigator Data Analysis Effect.* This category includes common errors in statistical methods.

a. The data analysis must be planned in advance to avoid a biased selection of data. As described earlier, the stopping and starting points of data-taking must be specified in advance.

b. If the data, after analysis, does not support the original hypothesis, the investigator sometimes makes a new hypothesis based on the data just taken. (This is called an *ad hoc* hypothesis.) He then claims that the existing data supports this new hypothesis. However, this procedure breaks the rule that the hypothesis must be made before taking the data and, in addition, is a form of circular reasoning. If a new hypothesis is made, the experiment must start over and more data must be taken.

c. The investigator must not attribute significance to random events. For example, suppose an investigator gives a number of tests and then announces proudly that 5 percent of the results are significant at the 0.05 level. But that is just what you would expect from chance. The importance of the discovery should not be exaggerated.

d. One possible cause of statistical bias is the fact that an investigator is more likely to check for computational errors when he obtains negative results than when he obtains positive results.

e. There is a deplorable tendency among some investigators to reject data that goes opposite to the expected effect. This is a subset of Investigator Fudging (see below).

f. Some parapsychologists go in the opposite direction and claim that negative results *prove* that ESP is taking place. In a telepathy experiment, getting *fewer* correct guesses than might be expected by chance is considered evidence that some paranormal influence is producing the result. This is a very convenient argument because accepting negative results as positive doubles the probability of getting an acceptible result in a given experiment. It is also a variation of the non-falsifiable argument by which you can prove anything you want.

5. *Investigator Fudging.* Fudging data consists of intentionally altering the results of an experiment. Cooking data, on the other hand, only changes it slightly to make it appear better. Some of our best scientists have been known to make their graphs look better by judiciously choosing the data. Even Isaac Newton is reputed to have engaged in such activity. It can amount to something as innocent as omitting a few outlying points on a graph so that the experimental points make a smoother curve. After all, if you are trying to prove that the acceleration is directly proportional to the applied force, you want to get something that looks approximately like a straight line.

Some investigators believe that it is legitimate to ignore data points that lie too far away from the average of all the data—say by three standard deviations—on the ground that these fluctuations are caused by unexplained experimental error. Other investigators consider such anomalies to be the real indicators that something interesting is happening. You accept or ignore these outlying points depending on what you are looking for—another example of investigator expectation cooking the data. Some statisticians think you should treat all points alike, unless you have a physical explanation for the fluctuation.

In the physical sciences, serious fudging is bound to lead to disaster for the investigator because somebody else is sure to come along and repeat the experiment. (See cold fusion, above.) In the softer sciences such as biology and psychology, where experiments deal with more variables and are harder to repeat, fudging may not be caught for a long time.

In parapsychology, fudging of only a small part of the data is needed to make the results of an experiment give a result in accordance with the wishes of the investigator. This is because quite small deviations from chance are considered satisfactory. Even J. B. Rhine complained vigorously about cheating by other experimenters in his laboratory. As late as 1974, and in spite of these repeated complaints, he found evidence of cheating by a colleague at the Institute for Parapsychology.

The temptation to cheat is in proportion to the rewards at stake. Often the rewards are not simply monetary, but take the form of a job, tenure, prestige, or simply vanity. Occasionally, the aim of a research is to obtain a "scientific" proof of one's own prejudices. Perhaps the most notorious case of fudging was the case of Sir Cyril Burt, a prominent psychologist who claimed that intelligence as measured by IQ tests was largely inherited. (Prejudice is involved because this kind of claim can be used to bolster notions of the inferiority of particular ethnic groups.) His procedure was to administer tests to pairs of identical twins who had been separated at birth; so they had the same inheritance but different environments. He then calculated correlation coefficients from the test scores of twins. (A correlation coefficient is a statistical measure of how closely pairs of test results agree with each other. A coefficient of zero means no agreement at all; a coefficient of one means perfect correlation.) The following table of his results shows astonishing consistency:

Year of test	Number of sets	Correlation
1955	21	0.771
1958	30	0.771
1966	53	0.771

Aside from the implausibility of the correlation being accurate to three decimal places (every physics teacher harangues his students about using too many significant digits), the absurdity of getting exactly the same number three times in a row is plain to see. When this improbable precision was noticed, the pathetic plot was exposed.

6. *Experimenter Personal Attribute Effect.* In studies that depend on personal interviews, the race, age, sex, etc., of the experimenter may influence the response of the subject. In a racially divided environment,

such as in the American South earlier in the century, blacks tended to tell whites whatever the blacks thought the whites wanted to hear. Early in her career, the famous anthropologist Margaret Mead was misled by Polynesians who told her myths about growing up in Samoa.

7. *Experimenter Failure to Follow the Procedure Effect.* The investigator may do his best to design a proper experiment, but if the people who actually do the data-taking or analysis do not follow the procedure exactly, error may follow. Indeed, even though the chief investigator plans his research with the best intentions, cheating by assistants can wreak havoc. This can happen through simple laziness, or because the assistant has his own agenda and wants the experiment to have a specific outcome.

8. *Experimenter Misrecording Effect.* In the past, when experimental results were written down by hand, errors could easily be introduced by inattention, carelessness, or unconscious bias. In a parapsychology experiment in which the subject's responses consist of a long list of random symbols, it is inevitable that some of the recorded data is going to be incorrect. Investigation has shown that the person recording the data has a tendency to err in favor of his own prior expectation of what the data should be. This effect is certainly less frequent now that automatic recording devices are universally used in research and the raw data, of course, can be stored for future review.

9. *Experimenter Fudging Effect.* The experimenter may fudge or cook the data just as easily, and for the same reasons, as his boss, the investigator. No more needs to be said on this subject.

10. *Experimenter Unintentional Expectancy Effect.* As we have indicated before, the expectations of the experimenter may be communicated to the subjects of the experiment in a number of ways. The experimenter may transmit his desires by unintentional paralinguistic cues (inflection or tone of voice) or by kinesic cues (body language or facial expression). In turn, the subject may respond to the experimenter by giving responses that will please the experimenter. The defense against this kind of error is the double-blind method, so that the experimenter cannot allow his expectations to color the outcome of the study.

There are so many pitfalls in the conduct of experiments involving human beings—especially when one is looking for small effects in a random sample—that no one should take any single piece of research too seriously. We see this every day when the news media present reports that "prove" certain foods or drugs to be dangerous, and then shortly thereafter cite other studies that demonstrate how the same foods or drugs are good for us. The first experiment must be considered a pilot study, which must be followed by repetition with bigger samples, better methods, and greater precision.

One thing psychologists agree on is that scientists are not objective, that they do their work with the same fallability as all human beings. Objectivity in science is obtained not by the individual scientist, who is always subjective, but by the scientific community as a whole. The examination of a scientific claim by peers with other points of view is the best way to grind rough edges off a theory and to detect any errors that may exist. The high precision of the instrumentation and the quick replication of an experiment possible in the physical sciences have made those sciences more objective as a whole than the "softer" sciences such as psychology. In the softer sciences, good experiments are difficult to do and the replication time is greater. But the ultimate aim of objectivity is the same.

NOTES

1. M. Rothman, "Cold Fusion: A Case History in Wishful Science," *Skeptical Inquirer* (Winter 1990): p. 161.

2. G. M. Miskelly, et al., "Analysis of the Published Calorimetric Evidence for Electrochemical Fusion of Deuterium in Palladium," *Science* (10 November 1989): p. 793.

3. R. I. Ewing, et al., "Negative Results and Positive Artifacts Observed in a Comprehensive Search for Neutrons from 'Cold Fusion' Using a Multidetector System Located Underground," *Fusion Technology* (November 1989): p. 404.

4. R. Rosenthal, *Experimenter Effects in Behavioral Research* (New York: Meredith Publishing Co., 1966).

5. D. Goleman, "Research Probes What the Mind Senses Unaware," *New York Times* (14 August 1990): p. C1.

6. I. Langmuir, "Pathological Science," *Physics Today* (October 1989): p. 36; M. Gardner, *Science: Good, Bad and Bogus* (Buffalo, N.Y.: Prometheus Books, 1981); T. Hines, *Pseudoscience and the Paranormal* (Buffalo, N.Y.: Prometheus Books, 1988).

7. T. X. Barber, *Pitfalls in Human Research* (New York: Pergamon Press, 1976).

8. T. S. Kuhn, *The Structure of Scientific Revolution* (Chicago: University of Chicago Press, 1962).

MYTH 12

"Scientists are always making false predictions."

False Positives

Scientists chronically put themselves into the position of making predictions about the future. It is an occupational hazard. Whether it is a warning of global disaster, or a prediction that cancer will be cured in the next century, the scientist puts his reputation on the line. If he errs, there is always a critic in the wings ready to crow about the wrongheadedness of scientists.

It's a risk that must be taken. Most predictions are chancy. Physicists can make precise forecasts of the future only under very special conditions. The prediction must be about a simple system, the initial conditions must be very well known, and there must be no effects that force the system to behave in a chaotic manner. If these conditions are in effect, the predictions can be quite certain; otherwise, the more complex the situation, the less certain the prognostication.

The same is true in other areas of science. Chemists can predict the outcome of reasonably simple reactions. An astronomer knows somewhat precisely where a space vehicle is going to be a year from now. The meteorologist, on the other hand, can barely predict what is going to happen tomorrow. Biologists, psychologists, and others working with human beings find it impossible to predict what a person is going to do within the next hour. Occasionally, however, they can make a generalized forecast with a high probability of being correct. For example, the chances are very good that I (and most of my readers) will be dead one hundred years from now. But that is a prediction of a statistical nature, based on national mortality tables. Similarly, a statistician can predict how many people will die each year out of a given group. Aside from being an expectation value—an average—it is also a projection. Projections are always of the form "if conditions remain the way they are now, then such and so will happen." Since conditions never remain the same, such projections rarely come true. Unexpected events such as war or fundamental improvements in medical science may change the statistics.

All of this means that most predictions of the future are going to be false; a few will turn out to be true, and some will be partially true.

In general, two kinds of predictions can be made: (1) the prediction that some specific event will take place, and (2) the prediction that some event will *not* take place. When we predict that something is going to happen, and it fails to happen, we call that a false positive. Similarly, a false negative occurs when we predict something is not going to happen, but it confounds us by happening.

The myth that scientists are always making bad predictions includes both kinds: false positives and false negatives. Like all the myths considered in this book, it is a false belief—or at least, a highly exaggerated one.

Often mentioned as an example of science falling on its face are early predictions of the great wonders to be derived from nuclear power. It is true that shortly after the atomic bomb made headlines in 1945, all kinds of marvelous future benefits were claimed for this new energy source: clean power to replace coal and oil generators, automobiles and airplanes with no need for refueling, nuclear-powered rockets for space travel, the utopian transformation of society by cheap energy, etc. None of these things have come to pass (at least not in the United States). Hence, according to critics, the scientists' predictions were wrong.

In the first place, it must be pointed out that many of these predictions were not made by scientists. No nuclear physicist in his right mind would predict the use of atomic power for propulsion of an automobile or civilian airplane. Knowledge of the neutron and gamma emission from a nuclear reactor would tell him that the shielding for a nuclear auto engine would have to be heavier than the automobile itself. The use of nuclear power for civilian transportation was probably the result of some magazine writer's flight of fancy during the heady days following the end of World War II. Even nuclear rocket propulsion has been dropped because of the risk of radioactive materials being spread throughout the atmosphere if an accident occurred.

As for the utopian transformation of society by atomic power, this myth can be traced back to a science-fiction writer—H. G. Wells—whose 1914 novel, *A World Set Free,* was the first depiction of re-emergent civilization after a nuclear holocaust. Wells's book was a remarkable bit of science-fictional prophecy. Starting with the recent (at that time) discovery of radioactivity, and recognizing the amount of energy locked up within the atom, Wells assumed that somehow this energy could be released with little cost and that the cheapness of this new energy would free the world from poverty and allow mankind to fulfill its promise. From that time on, atomic power has been a staple of science fiction. Reality impinged on the scene as soon as uranium fission was announced in 1939. The announcement was immediately followed by a number of stories touting the benefits and hazards of this new energy source. I wrote one in 1939, and Robert Heinlein wrote another in 1940.[1]

Fortunately, the real nuclear war, when it did come, was brief and one-sided. Unfortunately, civilian nuclear power did not produce utopia. Utopia was replaced by Dystopia.

The early predictions have been unfaithful to reality regarding the replacement of fossil fuels by nuclear power. A number of factors have derailed these prophecies. One is the less-than-perfect performance by some managers of nuclear power plants. The accident at Three Mile Island in 1979 destroyed much of the public confidence in nuclear power, resulting in the cancellation of all proposed nuclear plant construction in the United States. The deplorable safety record of Soviet nuclear power, especially as shown at Chernobyl in 1986, together with the truth emerging about conditions in American weapons plants, has eliminated what little public faith in nuclear power was left. What happened? It would appear that after legendary physicists such as Enrico Fermi completed the initial theoretical development of nuclear power, most of them went on to more sophisticated endeavors in nuclear and particle physics, leaving the design and administration of weapons factories and power plants to lesser scientists and engineers who proceeded to propagate a series of errors. The failure of prediction had more to do with human factors than with the inadequacies of technology.

However, early prophecies of atomic utopia were not totally implausible. Most of the American nuclear power plants have done their job without serious mishap. In France, three-quarters of electrical power comes from nuclear plants, and there has been no major accident to date. The reason, apparently, is good management, good education and training of the technicians who run the plants, and uniform design of the nuclear generators. It indicates that nuclear power has the capability of replacing other energy sources if properly managed. During the 21st century, as oil fields begin to run dry, the rest of the world may find itself forced to emulate France, and the original predictions of the nuclear scientists may eventually come true.

Will nuclear power be as cheap as predicted by some? Probably not, especially when all costs are accounted for, including the costs of storing nuclear wastes. One of the major flaws of science-fictional predictions is that costs are never a problem. Science fiction writers never have to meet a payroll or balance a budget.

Another prediction which has not as yet come true is that of fusion power. I do not mean cold fusion. I speak of good, old-fashioned thermonuclear fusion made by heating deuterium to a temperature so high that the nuclei would fuse, giving off energy. Back in the mid 1950s, when I started working in the field of fusion (at the Princeton Plasma Physics Laboratory, then known as Project Matterhorn), the standard prediction, made by reputable scientists, was that we would get the process working in twenty years. Twenty years later they were still saying, "just give us another twenty years." It got to be a folk saying about the lab that we would get fusion "twenty years from any given date." Actually, the twenty-year prediction was little more than a cheerleader's exhortation to keep up the spirits of the scientists and to keep the money rolling in from Washington. It was never a scientific prediction based on any kind of theory or evidence.

It is now nearly forty years since the national fusion effort (Project Sherwood) started, and the physicists in charge are claiming that they are finally in a position to create a self-sustaining fusion reaction which puts out more energy than is put into the machine. Unfortunately, the money to build the machine has vanished into the national deficit. We are finding that when it comes to building extremely complex systems in new fields of science, there is a great risk of underestimating the time and costs involved. The Manhattan Project was simple by comparison. Nature was most cooperative in bringing about nuclear fission. It is reluctant to cooperate in helping us get fusion. Therefore, we have to work harder. This is not to say that fusion power is a false prophecy. There is no physical reason for it not to work. The only questions concern our ability to make it work in a practical and economical manner.

Another potentially false prediction of the same nature concerns the superconducting levitating train. Theoretically it is a wonderful idea. I just don't want to be around when the liquid nitrogen springs a leak and the levitator ceases to be superconducting while the train is zipping along at 300 miles per hour. Safeguards will undoubtedly be put in place, but if it cannot be made fail–safe, it is a risky situation.

Fission power, fusion power, and levitating trains may or may not turn out to be false predictions. It is too soon to tell. They are examples of the fact that while truth is not always stranger than fiction, it is often much more complicated.

False Negatives

For many years, I have been carrying on a running argument with a well-known science fiction writer. He is the kind of person who does not like to admit that anything is impossible. Therefore, whenever I say something to the effect that faster-than-light travel is impossible, he rises to the bait with the hoary old chestnut: "Scientists predicted that man would never fly. They were wrong, weren't they? Therefore, you are just as wrong when you say we won't ever go faster than light."

The syllogism he seems to be making is:

a. All scientists are wrong when they predict something is impossible.
b. This scientist is predicting something is impossible.
c. Therefore, this scientist is wrong.

The fallacy of this syllogism is that its major premise is clearly wrong. Scientists are not always wrong when they say something is impossible. For one thing, they have never been wrong about perpetual motion machines, nor about the impossibility of levitation by psychic means. Therefore, the correct syllogism should be:

a. Some scientists are wrong when they predict something is impossible.

b. This scientist is predicting something is impossible.

c. Therefore, this scientist might be wrong or might be right. It depends on the kind of prediction.

Let us look at the facts concerning the prediction that man would never fly. Did all scientists make that claim? Clearly, no. Who did perpetrate the folly of saying that the flight of a heavier-than-air machine is impossible? That hapless person was Simon Newcomb, a mathematician and astronomer, head of the Nautical Almanac Office of the Naval Observatory, Washington, D.C., and, later, professor of mathematics and astronomy at Johns Hopkins University. In an article titled "Is the Airship Coming?", published in *McClure's* magazine, September, 1901, Newcomb predicted that the first successful flying machine "would be the handiwork of a watchmaker and would carry nothing heavier than an insect."[2]

Newcomb's words are a fine example of the need to be skeptical about the opinion of experts outside their areas of expertise. Of course, nobody was an expert in aerodynamics in those days. The more reason to be less dogmatic about the impossibility of air travel. Newcomb's views were shaped by the repeated failure of Samuel P. Langley to build a successful airplane.[3] The reason Newcomb's prediction was incorrect is that it was not based on fundamentals of physics. There is no law of nature that says heavier-than-air flying machines are impossible. It is simply a matter of getting the right materials together in the right configuration.

To give Newcomb his due, he had some good company. Lord Kelvin also thought that Langley was doomed to failure, as did the *New York Times,* which editorialized that Langley's experiments were a waste of public funds ($50,000, which in those days was real money). But in those days the *New York Times* had a poor batting average when it came to scientific judgement. Nine days before the Wright brothers' famous first flight in 1903, the *Times* insisted that man would not fly for another thousand years. But that was the the same paper which, on January 13, 1920, fulminated against Robert Goddard's belief that rockets could be used for propulsion in outer space. The *Times'* editorial writer seemed to think that rockets need air to push against. Another example of a non-expert out of his depth.

Certainly, it is not hard to find examples of those who claim one or another development in science is impossible. Some people have an unfortunate tendency to think that something is impossible simply because they do not know how the task can be accomplished. However, it is important to distinguish between those who do not see how a job can be done, and those who say the job is impossible because it violates one of the fundamental laws of nature.

In evaluating the value of a prediction, it is important to be aware of the two kinds of predictions mentioned in the previous section. The positive prediction says that something specific is going to happen in the future. The negative prediction says that some event is never going to take place. In another context, I have listed numerous reasons for the unreliability of positive predictions.[4] Among these reasons are system complexity, chaos, quantum uncertainty, and the inability of our brains and computers to handle all the variables involved. For these reasons, it is not surprising that most positive predictions turn out to be wrong.

However, there is one kind of positive prediction that is certain to be right. These are predictions that begin with the word "if". For example: if the world's population grows exponentially at a rate of 1 percent per year, then in 70 years the population will be twice its present size. Any person with a pocket calculator can then go on and compute how many years it will take to put one person on every square meter of space on the earth's surface. A prediction like this is a projection. It says "if things go on without a change of conditions, then certain consequences will take place."

The word "if" sets up special conditions and reduces the complexity of the situation. For example, when I predicted that all of us would be dead 100 years from now, there was in the background an implication: if no major changes take place in medicine during that time, then the prediction is true.

On the other hand, there is a different category of predictions that operate with no ifs, ands, or buts. These are predictions based on the laws of denial—those laws that tell us what kind of actions are not allowed to happen—perpetual motion machines, telepathy, UFOs hanging unsupported in the air, and all the other paranormal phenomena we are familiar with. Under Myth 3 (pp. 76f.) the connection was made between the laws of nature and the impossibility of the above paranormal phenomena.

It is important to realize and recognize that these laws are not "just theories" as claimed by some. These laws have been verified by the most precise experiments in science. While each law expresses one of the fundamental symmetries of nature, and thus is highly satisfactory from a theoretical point of view, the validity of the law rests on the experimental validation, and this validation is very good.

There is no case in the history of science where a scientist has been wrong in saying that a device would not work because it violated conservation of energy, conservation of momentum, or any of the other symmetry rules now recognized. This is a strong statement, but it is based on strong evidence.

We can see now why Newcomb was so wrong when he claimed that manned flight was impossible. He was relying on what was known then about aerodynamics, strength of materials, and Newton's laws of motion. This kind of knowledge is based on laws of permission, which tell us what kind of things might happen under given circumstances. We have seen that prediction with these rules is more hazardous than prediction using the laws of denial. In Newcomb's time, engineering mechanics was a relatively primitive science, and aerodynamics hardly existed. Why should anybody pay attention to dire warnings from somebody who was not an expert on flying machines? Newcomb's was a classical case of a little knowledge being a dangerous thing. Modern engineers, using computer modeling techniques, can come much closer to reality in predicting the feasibility of an aircraft design. Even so, testing of a physical model is needed to verify that the theoretical model actually works.

While predicting the precise behavior of a complex system can be difficult, what we can do is to set limits on allowable behavior. These limits are absolute. You simply cannot get past them no matter how hard you try or how ingeniously you scheme.

For example, present knowledge of solid-state physics enables us to make specific statements about the maximum strength of a solid material. We can calculate the maximum compressive force—the force that will crush the material—and we can obtain the maximum tensile force, the force that will pull the material apart. How can we be so sure about these limiting forces? We base the calculations on the fact that no material can withstand a force strong enough to separate one atom from another within the material's structure. This is a negative prediction, and is one more example of the fact that negative predictions are more certain than positive predictions. (In practice, the maximum force that can actually be withstood by a crystalline solid is less than the theoretical maximum, because solids always contain impurities and dislocations that interfere with the uniform crystal structure and so reduce the force required to break the crystal lattice.)

Here is another example of limits, involving one of the most important questions in electronics. The question is: how many transistors can you put into an integrated circuit of a given size? An integrated circuit is a network of transistors and other circuit elements inbedded in a chip of silicon or equivalent material. The modern computer is built around integrated circuits designed to act as signal processors or memories. The aim of designers since the invention of the transistor has been to reduce the size of each memory unit, since the smaller the transistor, the more memory units can be placed on a single chip. A side benefit is the increased speed of operation. At present, the greatest number of transistors attained per chip is in the millions. It is important to know the ultimate limit.

I can state with total certainty that nobody will ever make an integrated circuit out of silicon containing more than fifty thousand billion billion (5×10^{22}) transistors per cubic centimeter. We know that this number must be the absolute limit, because it is the number of atoms in a one-cubic-centimeter block of silicon, and it is simply impossible to make a transistor out of less than one atom. Actually, the true practical limit on the density of an integrated circuit is much less than the number given above, since it takes quite a few million atoms to make a transistor—the exact number being somewhat vague.

The most complicated system of all—the human brain, provides us with the most recalcitrant object of prediction. How can we predict what a person will do a year from now when the mind is so changeable? Styles, fashions, fads, morality change from one year to the next. (Take a look at any film from the 1950s to see that what was scandalous then is just ho-hum now.) But nature does not change its laws. As a result, what remains constant is the category of things that people *cannot* do because they are forbidden by the laws of nature.

Therefore, none of you in the audience will ever learn to levitate, unsupported by any natural force, several feet above the floor. None of you is going to make a fortune in the stock market by employing clairvoyance or by writing a horoscope. None of you is going to stay alive by eating one small, round pill a day instead of three square meals, and none of you is going to run a car by putting a fuel pill into a tank of water. None of you is going to be reincarnated into any form, no matter how good or bad you are.

It spoils some of the fun, but we must take nature the way we find it. There is nothing we can do to make nature do what it refuses to do. There is plenty of talk about "conquering nature" or "overcoming nature." But that's just talk. Nature only does what it does. We can't even talk about nature doing what it wants to do. Only humans want. Nature does.

NOTES

1. M. Rothman, "Heavy Planet," in *Astounding Science Fiction* (August 1939); R. Heinlein, "Blowups Happen," in *Astounding Science Fiction* (September 1940).

2. F. Howard, *Wilbur & Orville* (New York: Alfred A. Knopf, 1987).

3. I. Asimov, *Asimov's Biographical Encyclopedia of Science and Technology* (New York: Doubleday & Co., 1972), p. 585.

4. M. Rothman, *A Physicist's Guide to Skepticism* (Buffalo, N.Y.: Prometheus Books, 1988), chap. 5.

MYTH 13

"All problems can be solved by computer modeling."

Making Models

A new popular (and mindless) theme threads its way through the letters-to-the-editors columns. It is a proposal by animal-rights activists that medical researchers can obtain their desired results by computer modeling instead of experimenting on live animals. A careful analysis of this argument shows that it is simply another pseudoscientific notion that holds less water than a thimble.

The more extreme animal-rights activists are little more than the modern version of the old-time antivivisectionists. These are the people who earlier in the century disseminated pamphlets violently opposing experimentation on live animals—pamphlets replete with grisly photographs of dismembered creatures and horrible surgical instruments. Their appeal was more to our emotions than to reason.

The modern animal-rights enthusiasts make some attempt to reason with their audience. They claim that animals have just as many rights as humans. The term "speciesism"—the idea that higher species are superior to lower species—was coined by them as an analog to racism. This is a transparent bit of propaganda. The hope is that people of a liberal bent, through their commitment to equality and civil rights, will have a knee-jerk revulsion to the "elitist" idea that some animals are higher on the scale of evolution than others. The fact that all through evolutionary history some animals have preyed on others escapes them.

However, questions of equality or superiority or elitism as related to animal rights are ethical questions, not scientific questions. Scientists can be quite consistent in believing that they have an ethical right to use animals for the benefit of humans while at the same time making sure that the conditions of the experiment minimize pain and discomfort to the animal.

A separate question entirely—and one that does fall within the domain of science—is the claim made by the animal-rights propagandists that animal experimentation is no longer necessary for medical research, since we can now obtain the same results by means of computer modeling. This argument is a gross exaggeration of the capabilities of scientists. It is a fascinating paradox that a movement based on an antiscientific philosophy can have such a child-like faith in science's ability to find theoretical solutions to the most complex problems. On one hand, it displays an ignorance of scientific method. On the other hand, it suggests an awed confidence in the ability of an omnipotent father to heal all wounds and make a hurt go away.

This attitude bears some resemblance to a belief found in some quarters that science is "magic"—that scientists are magicians and can perform miracles. Indeed, the term "scientific miracle" is a cliche that crops up frequently in the media. The myth of the scientist as hero and as savior is seen here in full operation. (See Myth 11, pp. 161f.) Would that it were true!

The fact is that neither scientists nor their computers are omnipotent. Computer modeling is a useful technique for solving certain kinds of problems, and its use has steadily increased as computers have become faster and more powerful. However, every computer has its limitations, and these limitations are what make it impossible for computer modeling to solve every problem.

Precisely what do we mean by computer modeling? A model is a copy or imitation or representation of a real thing. We build ship models and put them in tanks of water to see how a real ship would operate. We photograph clothing on living models to demonstrate to buyers how they look on real people.

In science, a model might be nothing more than a set of mathematical equations whose variables represent quantities related to the behavior of a real system: the position and velocity of a group of particles, the electric and magnetic fields in a plasma, the flow of money in an economic system, for example. The number of equations depends on the complexity of the system and on the number of objects that have to be followed through space and time.

The path of the earth traveling around the sun can be represented by three equations: one for each of the three dimensions of space. Solving the equations permits the scientist to predict where the earth will be at any instant of time. However, three equations represents an oversimplified model. A truer model takes into account the motion of the sun as well as of the earth. There are then three equations for the sun and three for the earth—six equations in all.

Such scientific models permit us to visualize the physical appearance of a system and to predict how it is going to evolve in time (even if it is only an abstraction like a picture of a single atom). Unfortunately, a system that seems very simple may be represented by equations that have very complicated solutions. Just add Jupiter to the earth-sun system mentioned above and you get a set of equations which does not have an exact solution, except for a few special configurations. The problem of three planetary objects orbiting around each other under the influence of their mutual gravitational fields is a famous astronomical puzzle—the three-body problem—over which much sweat and tears have been shed. After centuries of effort by the best mathematicians, it was finally decided that the equations of the three-body system simply do not have general "analytical" solutions. That is, there are no solutions to the equations that can be represented by known mathematical expressions (such as sines or cosines or Bessel functions).

Nevertheless, if you want to predict where the three bodies are going to be at any instant of time, an answer can readily be obtained by a process known as numerical integration. The process works as follows. You begin by placing the three objects at a desired starting point and giving each a starting velocity. Then you write down the nine equations needed to describe the system—there are three bodies, and each moves in three dimensions. Since each of the three bodies influences each of the other bodies, the equations are coupled—they are simultaneous equations. (That is, the term that describes the force acting on each body contains the positions of the other two bodies, and so all the equations must be solved together.) The nine equations can now be used to calculate numerically where the three bodies will be after a very short time has elapsed. This procedure essentially ignores the change in the force while the bodies are moving the short distance from point one to point two. You now use this new position as a new starting point, and again compute where the three bodies will be a short time later. This process is continued for as long as you like.

It helps to have a computer to do the arithmetic because the shorter the time interval chosen, the more accurate the final result will be. But short time intervals require the use of more intervals to cover a given period of time, and you end up doing an enormous amount of arithmetic. (This is what is known by number crunching.)

Three bodies are relatively easy to calculate. Methods for numerical integrations are well known, and the programs are readily performed by even the smallest personal computer. But to do all the planets of the solar system requires thirty equations. The sun and nine planets make ten bodies, each of which moves in three dimensions. If you add the various planetary satellites, the number of equations grows to a respectable number, and major computing power is required to get an answer of reasonable accuracy for a model of the total solar system.

There are several levels of abstraction at work in the above description. The basic idea that the solar system can be described as a set of objects moving about through space, obeying Newton's laws of motion and the inverse-square law of gravitation, may be considered a conceptual model. It is the fundamental concept that makes the calculation possible. The set of equations that describes the system and the set of computer instructions that represents the equations is one part of the computer model. The other part of the computer model is the worked-out result of the number-crunching: the curves describing the planetary orbits as they change with time.

A model like this may be described as a bottom-up model. It starts with the simplest assumptions: nothing but the objects in motion and a knowledge of the forces by which they interact. It starts with a set of initial conditions (positions, velocities) which may be altered at will. From this starting point the computer proceeds to calculate how the system changes with time. Nothing else of a physical nature needs to be known.

Another kind of model is a top-down model. Here you start with large-scale objects and some knowledge of the laws by which they behave—for example, a weight hanging on a helical spring. If you know something about Hooke's law (the way the force exerted by the spring varies as the spring is stretched) and if you know Newton's second law (the relation between the weight's acceleration, its mass, and the force acting on it), then you can write a computer program that predicts how the weight is going to bob up and down on the end of the spring. You don't try reducing this model to interactions between the atoms in the spring; you work at a higher level of abstraction, represented by Hooke's law, which can be thought of as the summation of all the forces acting between the spring's atoms and the mass hanging from its end.

Both kinds of models—bottom-up and top-down—are important in the world of computer modeling, and we must investigate them in a little more detail before we can fully understand the implausibility of replacing animal experimentation with computer modeling.

Bottom-up Models

Drawing a picture of the solar system by setting up a computer model seems to be a simple, straightforward process. However, complications arise when the number of particles in the system grow large. The computer's ability to solve the problem depends on how many particles are in the system, how much memory the computer has, how fast it works (how many calculations per second it can do), and how much time is available for the computation. When I did my first three-body problem with an early model desk-top computer, it took about ten seconds to plot each orbit point for one of the bodies. Depending on how close together I wanted the points to be, it could take all day to complete one calculation. A late-model personal computer would run ten to twenty times faster. But for really large scale work, a supercomputer is needed, one that can follow hundreds of millions of commands per second.

To get a feel for the problem, consider that a popular personal computer, the IBM AT, can multiply 1000 by 1000 in about a thousandth of a second. This seems like a short time, but it makes for long waiting times if you have millions of computations to perform for each point on a plotted graph.

Why millions of computations? Consider a modeling problem that is important today. The problem has to do with nothing less than the large-scale structure of the universe. Astronomical surveys since 1986 have shown that the galaxies are not spread evenly through space, but that they tend to cluster in certain regions of the universe, forming strings and filaments, leaving enormous volumes—"bubbles"—devoid of visible matter. The size of these structures is so great that their extent is measured in millions of parsecs (Megaparsecs, or Mpc), a parsec being a little more than 3 light-years. One of these structures, called "The Great Wall" (discovered in 1986) is found to be 170 Mpc long, 60 Mpc wide, and only 5 Mpc thick.

The reason for the clustering of galaxies into clumps like this is somewhat mysterious. Analysis of the microwave radiation arriving at the earth from intergalactic space (the radiation left over from the initial big bang) shows that this radiation is uniform all through space to within one part out of 10^4. Why is matter not distributed uniformly to the same extent? Initial computer simulations of the universe, using galaxies as the fundamental particles orbiting under the influence of gravitational forces, failed to show any tendency to clump together, even when "dark matter" was added to the universe. (All theories of the universe require the presence in space of some kind of matter unseen to us in order to make the orbits of the stars within the galaxies and the motion of the galaxies themselves agree with what is observed.)

A more recent simulation was performed by R. Gott and C. Park at the Princeton University Department of Astrophysical Sciences, using a powerful computer to follow the motions of two million galaxies and two million parcels of cold dark matter.[1] Three simulations were performed, each requiring 14 hours of computer time to perform the billions of computations. The difference between this model and previous models lay only in the large number of galaxies employed. The striking result of the new calculation was that a model with a very large number of galaxies could bring out details of structure that the earlier computations, using smaller numbers, did not find. The conclusion drawn from the work was that the observed clusters and bubbles could be accounted for simply by a theory of gravitational instability. Starting with a uniformly spread-out mass of stars and galaxies, clumps would form by chance. These clumps would attract other clumps by gravitation, and after a long period of time there would be large-scale agglomeration of matter into dense clusters, which is the way the universe appears at present. The important thing is that no new and exotic forces were required to explain what is seen. In this calculation, four million objects were followed as they moved about under the influence of their mutual gravitational forces. It is clear that very large scale computational power was needed to handle this kind of problem.

The crucial point is that such a large number of objects *must* be used in order to get a useful result. Previous calculations with smaller numbers did not reveal the cooperative effects that come into play when very large numbers of particles act together.

This observation is very important: large numbers of particles acting together can create effects not visible when only a small number of particles are involved, even though nothing more than a single fundamental force is responsible for their interaction. It is this emergence of higher-level phenomena from more elementary phenomena, that provides all the interesting activities in the universe. Without this emergence of complex phenomena, the universe would be nothing but an amorphous cloud of dust. We will come back to this point many times.

The kind of calculation I have been discussing is a bottom-up model. It is called bottom-up because we start at the bottom of the system with simple particles moving about under the influence of one or two well-known forces, and work our way up, trying to predict what those particles are going to do. The particles might be atoms, molecules, planets, or—as in the example given above—galaxies. The reason a galaxy can be treated as a particle is that the distances between galaxies are large compared to the size of a single galaxy; furthermore, in this calculation, we are not interested in what happens inside a galaxy. In theory, starting from the bottom, we should be able to deduce all the actions taken and all the structures formed by any complex system. In practice, there are enormous obstacles in the way.

Consider the apparently simple problem of modeling a container of an ordinary gas such as hydrogen. To be specific, let us take one mole (one gram-molecular weight) of gas as our sample. We know that it occupies a volume of 22.4 liters (about six gallons) at room temperature and pressure. We know that the gas consists of nothing but independent molecules bouncing about like billiard balls, and attracting each other weakly if they get very close together. From this very simple picture we should be able to predict all the properties of this gas and everything that it does.

However, if we try to do a brute-force computer model of this gas, we immediately run up against an impregnable stone wall, for the modest volume of gas chosen contains altogether 6×10^{23} molecules. There is no way for any existing computer to follow the individual paths of all the molecules in the container. This possibility does not exist simply because the number of molecules in the system being studied is of the same order of magnitude as the number of atoms in the computer! There is not enough memory in the computer and there is not enough time to do the calculations. If the computer could do one billion calculations per second, it would require 18 million years to follow all the molecules through one step of motion! When dealing with atoms or molecules *en masse,* more sophisticated methods are required.

There is a way to get at least an approximation to the truth. What we do is divide the volume of gas into a number of small parcels. Each of these small parcels contains a large number of molecules traveling in different directions at different velocities. We then average over the velocities of the molecules within each parcel and emerge with a set of equations that describes the motion of each parcel as a whole. These equations turn out to be none other than the well-known equations of hydrodynamics, which relate the motions of the gas parcels to variations in temperature and pressure. These are complicated equations and, in general, are not solvable in an exact way. However, by making various approximations, one can obtain simpler equations that are solvable. For example, leaving out the time factor gives us the static equation for the universal gas law, which relates the temperature and pressure to the volume of gas under constant conditions. Or, leaving the time in to a first approximation, we obtain the equations for sound waves in the gas.

Notice that properties such as temperature, or pressure, or wavelengths do not appear in the original model. They are not properties of the individual molecules. They are *emergent properties,* i.e., they emerge from the averaging process that has been applied to the model. The temperature is the average kinetic energy of the molecules in each parcel. The pressure is the average force per unit area exerted by the molecules of one parcel on the molecules of a neighboring parcel. Wavelengths can be observed only when we look at the perturbations in a volume of gas. They relate to the changes in the average positions of the molecules, or changes in the gas pressure at each location.

These emergent properties are, in a sense, high-level abstractions, appearing as results of thought processes that we apply to the basic system. Each emergent property requires a human definition. The lowest level of abstraction consists of nothing more than a multitude of molecules moving about in a container. When we look at them at that low level, we see nothing but chaos. But when we pull back and visualize them at a higher level, we see the patterns: the flows, the waves, the transmission of energy and information. However, these phenomena are not merely mental images. Wavelengths, frequencies, amplitudes, temperatures, and pressures are physical properties of matter that can be measured with the help of appropriate instruments. They are real structures and configurations in the fluid. They may be permanent, or they may exist temporarily and dynamically in a changeable medium.

Instead of limiting ourselves to an ordinary gas, let us see what happens when we apply the same procedure to an ionized gas, obtained by separating the atomic electrons from their nuclei. The gas is then called a plasma and consists of positive and negative electric charges floating about. A magnetic field can be applied from the outside to guide the motions of the particles and confine the plasma within a definite volume of space. We now do the same trick we tried before: breaking the plasma up into little parcels and averaging over the velocities. Doing this, we obtain the equations describing the magnetohydrodynamics of a plasma immersed in electric and magnetic fields.[2] These are formidable equations, for the motions of the charged particles may produce their own electromagnetic fields which in turn influence the motions of the charged particles. The result is a feedback system that can produce a great array of waves and instabilities. The hydrodynamic equations describing the system can be solved directly only in the simplest cases. However, a computer model of the plasma can be constructed by making the little parcels the elements of the model. Choosing a number of elements small enough to be calculated, you can see all the behavior of the plasma emerging, complete with hydrodynamic flow, waves, instabilities, turbulences, etc. You can also see that certain types of waves can be used to heat the plasma, while certain instabilities cause the plasma to escape from its confinement in the magnetic field.

This is the kind of computer modeling that has been of great value in thermonuclear fusion research, since that research invariably involves the heating of a plasma to very high temperatures while it is confined for a time long enough for energy-producing nuclear reactions to take place.

Can we apply the principles of computer modeling to liquids and solids to predict the structure of large molecules and their organization into living matter?

Clearly, the large number of atoms in any structure visible to the naked eye precludes a brute-force approach. It is possible to use the fundamental equations of quantum mechanics to predict the structure of molecules containing several atoms, and the results have been very satisfactory to date. However, when dealing with large-scale objects, it is necessary to substitute cleverness for number-crunching.

One trick is to eliminate those aspects of a system which do not contribute to the behavior you are looking for, and to concentrate on only those aspects that cause the macroscopic effects of interest. In an effort to understand the behavior of water near or below the freezing point, a group at Boston University recently performed a computer simulation which treated liquid water as a random network of hydrogen bonds linking water molecules over long distances.[3] (A hydrogen bond arises from the fact that water molecules are dipoles: they are positively charged at one end, and negatively charged at the other end. The hydrogen bond is the electric attraction between these dipoles.) These bonds break and reform over times measured in picoseconds (a millionth of a microsecond). At any given time there are enough intact bonds to give water its structure.

The computer simulation tracked 216 water molecules in a cubic box 18.6 Angstroms on a side. (One Angstrom is 10^{-10} meters.) Computations were done at five different temperatures; the aim of each calculation was to determine how long each bond lasted. The simulation showed that the bonds do not suddenly break off at a fixed time, as earlier theories had assumed, but that there is a continuous distribution of times as a function of the energy required to break the bond. The average lifetime is found to increase rapidly as the temperature decreases. From this knowledge, the researchers hope to explain why the behavior of water becomes so strange at temperatures near its freezing point. (Water expands as it freezes, in contrast to most other materials.)

In this case, use of a moderate number of molecules was sufficient to bring out the necessary information. The reason for this is that the experimenters were not trying to simulate structures within the system, but only to investigate the statistical nature of a simple system—to find how some average quantities such as the lifetime of the chemical bond varied with temperature. It is clear, on the other hand, that to model the behavior of complex systems such as a living cell, or the earth's atmosphere, other approaches are necessary.

These other approaches—included in top-down methods—require us to know something about the large-scale structures of matter and to make models simulating the way these structures interact with each other.

Top-down Models

As the name implies, a top-down model is one which starts at the top with a large-scale structure and tries to describe its behavior in terms of high-level laws. This is the kind of model that must be used if you are building a machine and you would like to predict how this machine is going to work. Or perhaps you are designing a chemical manufacturing plant and you must calculate the conditions needed for the most efficient production. To accomplish this end you start with a description of the system, plus a knowledge of the chemical reactions by which it operates. Put this information into the computer, grind through the calculations, and theoretically the computer tells you what the plant is going to do. You can then vary the operating conditions—temperature, pressure, chemical concentrations, etc.—and look for the best combination.

For this method to be effective, the computer model must meticulously simulate the real device in all its details. The difficulty is that sometimes you don't know ahead of time what details are going to be important. Consider the weight bobbing up and down at the end of a spring mentioned earlier in this chapter. This device is studied by every freshman physics student. It is an exercise in using Hooke's law and Newton's law of motion to predict the behavior of a simple mechanical system. Hooke's law says that when you stretch a spring, it pulls back with a force proportional to the distance stretched. Double the distance and you double the force (or vice versa). This is by no means a fundamental law of nature. It is simply a rule based on the observed fact that elastic materials behave this way. In fact, this is just what we mean by an elastic substance in physics: it is a material which obeys Hooke's law.

If you have nothing more than a mass bobbing passively on the end of a spring, the solution to the problem is very simple: the motion is a free oscillation, represented by a sine wave, the frequency of which can be calculated without resorting to computer simulation. However, if an oscillatory force is applied to the spring (by moving the top of the spring up and down), then the system is in a state of forced vibration, and the equation describes how a resonance takes place when the frequency of the driving force equals the frequency of free oscillation: the amplitude of oscillation is a maximum at the resonant frequency.

Occasionally, however, things happen that you don't expect, things not predicted by the simple theory. If you take away the driving force, you would think that the oscillation should go on forever. Instead, it gradually dies down, even when you get rid of air resistance by putting the device in a vacuum jar. Or if you increase the driving force too much, you might find resonances at two and three times the fundamental frequency.

These effects are hints that the simple equations don't tell you everything that is going on. They don't say anything about internal friction; rubbing of the spring particles against each other, causing the motion to dissipate. And they don't say anything about what happens when you get past the elastic limit—that is, when you stretch the spring far enough so that the stretch is no longer proportional to the applied force and you get into the realm of non-linear effects. These effects demonstrate the fact that Hooke's law is not a fundamental law: it is a top-down law and is simply a description of observed macroscopic behavior. It applies only within strict limits—as long as you don't apply too much stretch—and it applies in a slightly different way for each kind of material. Some materials do not obey it at all (taffy or lead, for example).

All of this is to illustrate that anybody trying to make an accurate computer model must be certain that he is entering into the computer all the necessary details concerning all the parts of the structure, as well as all the laws governing their behavior. If you use approximations, you must be sure that the device is operating in a domain where the approximations apply. (Hooke's law is but a first-order approximation to the truth, good up to the elastic limit.) If the experimenter does not do this, then the computer is going to present him with a host of unpleasant and unexpected results. These results may or may not represent what actually happens. One thing is certain: you will never be perfectly confident about the computer simulation unless you try it out on a real model built with real materials. (Simon Newcomb's prediction about man never flying came a cropper because his model did not apply.)

Another kind of model is needed when you begin with a very complex structure—say the human brain. Knowing something about the behavior of the brain—what it does in response to stimuli, for example —you hope to deduce something about what goes on inside the brain at a lower level.

Unfortunately, the brain has so many inputs and outputs that it is hopeless to deal with it simply as a black box that can be analyzed by observing how the outputs vary in response to changes in the inputs. The psychologist may observe the behavior that each stimulus evokes, but what he learns mainly is behavior, not what goes on inside the box. To learn what goes on at a lower level the neurologist probes the neurons in the nervous system, observes the kinds of signals they generate, and tries to see how they interact with each other. While this technique is bringing neuroscientists closer to the goal of understanding how people see, learning how people think is a more distant goal.

An approach that seems to hold some promise is not quite a top-down approach. We might describe it as a "below-the-top method". The starting point here is at a level just below the surface. The scientist uses whatever he knows about the nervous system and the structures that transmit and manipulate signals. He then tries to picture what kind of electronic circuitry might be used to simulate the structures in the brain. The next step is to build such circuits and see what they do when they are turned on. If the result of this attack resembles in any way the behavior of the human or animal (or insect) brain, then an advance has been made.

In recent years, a type of circuit called a neural net has been the object of extensive experimentation, and the results are considered hopeful by many. This method uses simple electronic circuits, each one of which simulates the behavior of a single neuron. Each elementary circuit has several inputs and one or more outputs. Electrical pulses are fed to one or more inputs. A pulse coming to one of the inputs may excite an output pulse, or it may inhibit an output pulse, depending on the nature of the circuit. It may take two or three inputs to excite an output. To this extent the device resembles the coincidence or anticoincidence circuits familiar in physics laboratories. A network of such elementary circuits connected together may perform a variety of operations such as pattern recognition.[4]

The original neural nets made in 1941 by McCullough and Pitts were capable of only a limited repertoire of activities. During the 1970s, Grossberg and Hopfield expanded the repertoire by allowing the pulses to take on a variety of amplitudes instead of just the two (on and off) used by the original nets. Networks of this kind turn out to be trainable: they can learn behavior and languages and come closer to mimicking human behavior than other types of computer programs. One such network is presently being used to pilot a car through a prescribed route.

The obvious question is: how far can neural nets go? Will they ever imitate human behavior in all ways? Probably not, at least not in their present form. No matter how complicated we make neural nets, they will not imitate human behaivor in all details. There are at least two reasons for this.

First, the human nervous system is more than simply a network of neurons. It operates in a chemical environment; the pulses traveling through the nerves are electrochemical, not purely electrical. In addition, neurotransmitters such as norepinephrine, serotonin, and dopamine travel for long distances from one part of the nervous system to the other to affect the functioning of the synapses. They seem necessary for the proper functioning of the whole person; you don't feel right if they are not present in the proper amounts. Perhaps they are necessary for feelings and sensations.

Second, human beings do not operate in a vacuum, but require programming over a period of many years. This is the training applied by parents, by neighbors, schools, and the rest of the environment. Even if a computer could be built with human-like hardware, it would require human-like programming and human-like software in order to function like a human.

All of this is to illustrate why the idea of replacing animal experimentation by computer simulation is largely wishful thinking. Living beings are simply too complex to be modeled with the precision required for useful results. Suppose, for example, that pharmacologists had a computer list of all known drugs, complete with molecular structure. Using computer modeling, they could determine how each drug would fit into the molecular structure of a virus and so could predict whether the drug would inhibit the multiplication of this virus. At first glance, this method would seem to eliminate the need to test the drugs on animals.

As a matter of fact, such a scheme has actually been used to seek a drug capable of fighting the HIV virus that causes AIDS. A data base containing structural images of thousands of known drugs was searched to find a molecule with just the right shape to bind and inhibit the activity of a key HIV enzyme. The drug Haloperidol turned up as the best fit. The good news is that the drug did work when tried on animals in the laboratory. The bad news is that when the drug works, it only does so at doses high enough to kill the patient.[5]

The moral is that the computer modeling presently feasible does not tell the scientist whether the drug will actually work in an animal or human being, whether the dose required is less than the lethal dose, what the side effects are, etc. Testing the drug on a living thing is still necessary. The computer can only tell you whether the drug might be useful; it does not say that it is useful. Furthermore, it says nothing about new and unknown drugs that are not already in its list.

When faced with the absurdity of their claims for computer simulation, animal-rights activists often fall back on their last line of defense by insisting that medical advances have never depended on animal experimentation. This is a vicious distortion of history and is easily refuted by medical scientists who are not aware of any advance in medical science that did not require animal experimentation at some stage of the process.

Here is one major example of personal interest to tens of millions of people in this country alone. In 1921, Frederick Banting demonstrated, by using dogs, that the Islets of Langerhans in the pancreas generate the hormone insulin necessary for the metabolism of glucose in the body. Before this work, millions of diabetics were doomed to an early death. The animal-rights people may claim that this work could have been done without animals, but they would be hard put to show how. In 1921, there were no computers.

NOTES

1. B. Schwartzschild, "Gigantic Structures Challenge Standard View of Cosmic Evolution," *Physics Today* (June 1990): p. 20; C. Park, "Large N-Body Simulations of a Universe Dominated by Cold Dark Matter," *Monthly Notices of the Royal*

Astronomical Society 242 (1 February 1990), Short Communications, p. 59.

2. L. Spitzer, *Physics of Fully Ionized Gases,* 2nd ed. (New York: John Wiley & Sons, 1962), chap. 2 and appendix.

3. I. Peterson, "Computing Liquid Water as a Transient Gel," *Science News* (14 April 1990): p. 231; S. Sciortino, P. H. Poole, H. E. Stanley, and S. Havlin, "Lifetime of the Bond Network and Gel-Like Anomalies in Supercooled Water," *Physical Review Letters* (2 April 1990): p. 1686.

4. A. K. Dewdney, *The Turing Omnibus* (Rockville, Md.: The Computer Science Press, 1989), chap. 24.

5. *Science News* (23 June 1990): p. 390.

MYTH 14

"More technology will solve all problems."

Limits to Growth

Two decades ago, a remarkable book burst upon the world of scholarship. The name of the book was *The Limits to Growth: A Report for THE CLUB OF ROME'S Project on the Predicament of Mankind.*[1] Its aim was to explore the consequences of exponential growth in all the activities of mankind: population expansion, use of energy, mining of minerals, pollution, disposal of waste products, etc.

Why *exponential* growth? It is a fact that all kinds of animal or plant populations grow in such a way that the increase in population during a given year is a certain percentage of the population present at the beginning of the year. For example, imagine that during each year one out of every hundred people in this country had one child (over and above the number of deaths). It is clear that the population of the country would increase by one percent during that year. What would happen, then, if the population continued to grow at a rate of one percent per year? What would happen is that for every million people at the beginning of the first year, the end of that year would see an increase of 10,000 people. The million has grown to 1,010,000 in one year. The next year finds an increase of one percent of 1,010,000, which is 10,100. We see that even if the growth *rate* remains the same, the numerical increase during the year gets progressively greater because the population at the beginning of the year has increased over the year before. (Of course, nothing guarantees that the growth rate will remain fixed, but that is another problem.)

You recognize that we have been describing the development of a compound interest curve. It is the same kind of increase you would get if you put money into a bank at a fixed rate and just let the money accumulate. What characterizes this kind of curve is the fact that the population (or the amount of money) always doubles in a fixed period of time. The doubling time depends on the growth rate. A population which increases at a one percent rate per year (as we noted under Myth 3, p. 70) is going to double every 70 years, which is only a moment in history. (To be exact, it is closer to 69.315 years, but let us use 70 years as an approximation.) In the same manner, money accumulating in the bank at a 10% interest rate is going to double every seven years. That is a very interesting figure for bank depositors. On the other hand, an inflation rate of "only" five percent—which everybody accepts as normal these days—guarantees that your savings are going to be worth half their present value in fourteen years. That is an example of a negative growth rate, and a fact that all retirees should be aware of.

Of course all of this is (or should be) familiar to any mathematics student. What The Club of Rome did was to assemble a group of experts in various fields such as population, pollution, resources, etc., to construct a computer model intended to predict how the world, considered as a holistic system, would behave in the future under certain assumptions of population growth, energy use, material resources, pollution production, etc. In a simplified way, it pictured the connections between numerous demographic and economic variables: if population grew at a given rate, and if each person used up a certain quantity of resources and bought so many cars and drove x miles per year, and produced so much waste material to be disposed of—then these resources would be used up by a certain time, the garbage dumps would be filled up by a certain time, the oil deposits would be used up by a certain time, and civilization would be in trouble.

In brief, what the computer model predicted was what one might expect from simple commonsensical logic: if population increases at an exponential rate, then any finite, non-renewable resource within the earth must eventually be used up. There is no way to argue oneself out of this conclusion. The only question is: how much time is this going to take? How much time the world has left depends on the rate of population growth, the energy and material use of each individual, the total energy and mineral resources in the earth, and so on.

A separate question is this: can the world reach a reasonable and gradual equilibrium by stabilizing the population, or must there be a sudden and total collapse of civilization caused by population outracing the resources of the world?

To those of us who read *Limits to Growth* and perceived all of these conclusions to be a necessary consequence of the mathematics, it was astonishing to read the objections that appeared in the newspapers and journals. Many of the objections were technical, claiming that the computer model used by The Club of Rome was oversimplified and left out important details. There seemed to be no comprehension of the fact that the essence of the book did not lie in the details of the computer model.

The point of the book was startlingly simple: it was an exercise in imagining what happens when things grow at an exponential rate without limits. Even if the growth is not exponential, the one inarguable conclusion is this: if you use up finite resources at an increasing rate, you are going to run out of these resources sooner or later. As a matter of fact, you will eventually run out of these resources even if you use them up at a constant rate. It's only a question of when it happens. Exponential growth is particularly insidious because of the apparently sudden manner in which the end comes.

In its first chapter, the authors of *Limits to Growth* quote a French children's riddle to illustrate the abruptness which which things disappear when they are used up in an exponential manner. Consider a pond on which a water lily is growing. It grows in such a way that its area doubles every day. If allowed to grow unchecked, it will fill the pond completely in thirty days. For the first few days, the plant is quite small and you decide there is nothing to worry about for a long time. At fifteen days, you notice that the plant is beginning to spread out a bit and you make up your mind to cut it back when half the pond is covered by the water lily. Now comes the riddle: *on what day do you have to use the clippers?*

This is a problem that would make most students leap for their calculators. But if you stop to apply logic instead of brute-force arithmetic, the problem can be solved easily and instantly in your head. The trick is simply to work backwards. *If the pond is full on the thirtieth day, then on the twenty-ninth day it had to be half full.* (This is true because we said the area of the plant doubles every day.) It seems as though on the last day the water lily suddenly explodes into the remaining half of the pond. And then you ask: why didn't we see this coming?

Most astonishing are the number of people who continue to object to the Limits to Growth concept. In particular they object to the necessity of limiting population and conserving resources. Some of them say: we will always find a new way of obtaining resources, and new ways of generating power. Some others say: there is an indomitable historical progression; we started out burning wood, then coal, then oil, then uranium. Something will surely be discovered in the future to save us when we use up all other fuels. Some of an extremist bent speak of the infinitude of Space. We can always go to the other planets. Never mind the cost. Never mind the fact that the other planets are not hospitable to life. New technological advances will enable us to overcome all of these difficulties.

Here we arrive at the new myth: modern technology will solve all our problems.

At this point the debate begins to take on theological overtones. At least two kinds of theology appear to be involved:

First, there are the economic conservatives to whom *growth* is a philosophical, if not theological, necessity to the maintenance of "civilization" (or to the maintenance of their lifestyles in the manner to which they have become accustomed).

Second, there are those to whom the phrase "population control" means birth control, a practice forbidden by their religions. It is easy to see why limits to growth are unthinkable to those whose minds have been conditioned from birth against limiting conception.

In addition, there are a number of understandable nationalistic agendas to contend with. While it is easy for those in developed countries who have been using most of the world's oil to say it's time to level off, the majority of the world wants to catch up, to become industrialized. But if all countries use oil at the same rate as the United States, the world's energy supplies will certainly not last beyond the first half of the twenty-first century. A serious and intractable dilemma faces us.

The Inadequacy of Technological Fixes

There are two common responses to the problem of diminishing energy resources—in particular to the coming shortage of oil: (1) we make use of energy conservation and build more efficient machines, and (2) we find alternative, renewable energy sources. These are desirable goals for the short term, and it is not my intention to criticize them. Unfortunately, in the long run these actions will turn out to be sadly inadequate as long as the population of the world continues to increase exponentially.

To put it another way: *no technological fix is going to have more than a temporary effect on diminishing resources unless the growth of population is halted.*

A simple example illustrates why this is true. Let us suppose that in the near future the thermonuclear fusion research program finally meets with success, so that deuterium can be used as fuel for generating electrical power. Deuterium is an isotope of hydrogen that is found in sea water, and while it makes up only 0.015% of the total hydrogen in the water, there is so much water on earth that the potential amount of fuel is very great. Let us suppose, for the purpose of illustration, that if the world population stays constant and continues to use energy at the present rate, there is enough deuterium in the sea to last a million years. (Most estimates give a much greater number of years, but the exact number has no effect on the overall argument.) A simple calculation shows that if the world population rises at a uniform rate of one percent per year, and if each person continues to use the same amount of energy as before, then this million-year stock of fuel will be gobbled up in only 920 years.

The reduction of time from a million years to 920 years illustrates the enormous power of exponential growth. Although one percent per year does not seem like a great rate, we saw in the last section that it means a doubling of population every 70 years. Doubling and redoubling for 920 years means a huge increase of population. (We saw in Myth 3.3 that a 1200-year growth would put one human being on each square meter of the earth's surface. That's why it makes no difference whether there is a one million or a ten thousand million year supply of deuterium in the sea.)

Now let us ask what happens if we double the efficiency of all our energy-using contrivances so that each person uses only half as much energy as before. How far does this improvement stretch out the fuel supply? Remember that the population is still increasing at the same rate. All we are doing is reducing the energy consumption per capita, which is what everybody is now trying to do. After an appropriate calculation, we obtain a rather disappointing result: we find that the fuel in the ocean is now going to last for 990 years. From 920 to 990 does not seem to be a major improvement. In fact, we notice that the difference is only seventy years. Is it a coincidence that this is the same as the time it takes the population to double?

Not at all. There is a definite mathematical reason for the fact that the increase in the fuel's lifetime is the same as the population doubling time. A graph showing the amount of fuel consumed annually is seen to be a curve that rises with a slope that depends on the population growth rate. In this case, it is an exponential curve doubling its height every seventy years. This kind of curve becomes a straight line when plotted on a logarithmic scale, making it easy to see what happens. In Fig. 14-1 (p. 211), we see the fuel consumption curve based on a one percent per year population growth rate. In this figure we see two lines. One is based on each person using one unit of fuel per year. The other line depicts what happens when every person reduces his or her consumption by 50 percent. Each year the height of the second curve is half the height of the first. However, since the population increase is the same for both curves, the total consumption rises at the same rate. The consequence is that seventy years after the starting point, the height of the second curve has doubled and is now at the point where the first curve started. The net result is that the second line looks just like the first line, only delayed by seventy years.

What makes the increased fuel efficiency so inadequate in prolonging the fuel lifetime is the inexorable doubling of the population every seventy years. This conclusion holds for any kind of finite resource on earth, whether it be living room, mineral deposits, or space for disposal of wastes.

This proposition can be generalized. Suppose population increases exponentially with a doubling time of n years. If each member of that population cuts in half the use of any fixed resource, then the lifetime of that resource will be extended by n years.

We see that conservation of resources, while of temporary value,

EXPONENTIAL GROWTH OF FUEL USE

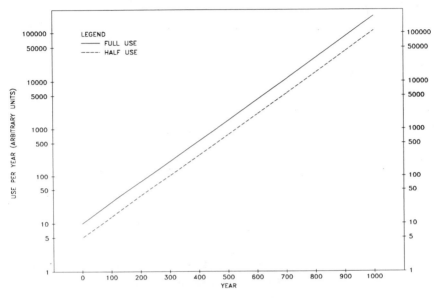

fails in the long run as long as the population is allowed to increase exponentially. In the course of history, seventy years is only an instant, and while a thousand years is a reasonably long time, it is short compared with, say, the life span of Egyptian civilization. If we are responsible people, we should like our descendants a thousand years down the road to have a few resources left.

Therefore, we must consider the ultimate solution to the problem: reducing the population growth rate. To show how powerful is the effect of reducing growth, we return to our example of deuterium fuel and ask what happens if we manage to cut the population growth rate in half—from one percent to one-half percent—while keeping the per capita use of energy unchanged. The result is impressive. A modest application of population control increases the lifetime of the deuterium from 920 years to 1700 years. This is a more satisfactory improvement—780 years instead of only 70 years. Even so, on an extended time scale the maintenance of civilization for another 1700 years is only a brief blip on the chart. We already have 6000 years of recorded history. We would like it to at least double.

There is only one way to do this and that is to cut the growth rate of the earth's entire population to zero. This must be done before any technological improvements can extend our resources more than a few hundred or a few thousand years. Yet this is a solution barely mentioned in the media at large. Typically, long articles which speak of the need to conserve fuel, or to use renewable resources such as solar power, devote perhaps a single timid line to the need for ending population growth. Only organizations actively dedicated to the need for limiting population expansion—Zero Population Growth or the Worldwatch Institute—are totally aware of the gravity of this problem.

Indeed, the United States government policy seems to be aimed in the opposite direction. During the 1980s, the Reagan administration made it illegal to provide funds for family planning (the code term for birth control) to countries which allow abortions. Since it is impossible to defend such a policy on pragmatic grounds, one can only conclude that some of our government officials are guided by a philosophy of supernatural beliefs rather than by a philosophy of realism.

When we investigate world population trends as they are in reality, we find that the situation is more complicated than the simple examples given above would indicate. Many of the industrialized nations actually have reduced their growth rates to close to zero, showing that they have indeed paid some attention to the problem. A number of European countries have achieved growth rates of less than 0.2 percent per year. Japan had a growth rate of 0.7 percent per year, but is taking steps that should reduce the growth to zero by the end of the twentieth century.[2] In the United States, the greatest contribution to population growth is immigration; this must be stabilized if zero population growth can be attained. But, of course, that means that the poorer countries from which the immigration originates must also slow down their growth rate.

In the Middle East, Africa, and Latin America, growth rates of two to three percent are common. Most demographers project that the world population will continue growing until it levels off at some 10 billion, and that most of the 5 billion additional lives will be concentrated in the lesser developed countries, where poverty and lack of education result in high growth rates. A three percent rate represents a doubling time in the vicinity of twenty-three years. Such growth is especially devasting in countries where the gross national product grows at an even slower rate. Countries of this nature, mainly in Asia and Africa, have little chance in sharing the industrial revolution.

Let no one think that the imaginary growth rates I have been playing with in this section can be maintained for any great length of time. In countries with growth rates of 3 percent, the individual person becomes poorer over the course of time. (Unless there are extraordinary circumstances such as the presence of copious oil reserves to enrich the population for a few years.) Consequently, disease and death take over, the life expectancy of the population suffers, and the growth rate goes down. Thus, unlimited growth is self-limiting—but at the cost of civilization itself.

Renewable Sources of Energy

A popular rallying cry when fossil fuels run short is to demand the development of alternative, renewable energy sources: solar power, alcohol, biomass, etc. All of these are fine ambitions and are worth pursuing. However, just as in the case of thermonuclear power discussed in the last section, nothing helps in the long run if you don't stop exponential population growth.

Solar power provides a salutary example. The sun constantly deposits on the earth's surface about 4×10^{16} watts of power. Let us be optimistic and suppose that we are able to use half of this power. An efficiency of 50 percent is higher than anything I have seen in the literature, but this is only a hypothetical example, and I want to err in a direction that makes things look better than they actually are.

The total power consumption in the world is currently somewhere in the neighborhood of 10^{12} watts. We now ask the question: if we keep using power at the same rate per capita, but the population increases at a rate of 1 percent per year, how many years will it take before we have to cover the entire planet with solar collectors in order to provide everybody with energy?

First, notice that the total power available from the sun is about 20,000 times greater than the power being used now from all sources. (Note that when we burn fossil fuels, we are using their energy at a much greater rate than this energy was originally stored in the fuel from the sunlight.) Thus, even if we are forced to give up fossil fuels and switch to solar power, we can increase our power consumption 20,000 times before the entire planet is enclosed in power collectors.

Running the numbers through the calculators, it turns out that power use must be doubled only a little more than 14 times to reach that increase of 20,000. And with a doubling time of 70 years, that comes to just 1,000 years. So the most we can do with an increase of non-renewable resources is to buy a thousand years of time, unless the population stops expanding at the present rate.

Again, we see the enormous power of exponential growth. Remember that what I am presenting to you is a totally hypothetical numbers game. It is based on certain clear assumptions that are stated up front. Certainly, the assumption of one percent world population growth is going to break down before a few more decades have passed. If birth control does not do the job, then the job will be done by an increase in the death rate, whether we want it or not.

When the Technological Fix Makes the Problem Worse

It is by now a commonplace that nuclear energy has caused more problems than it has solved. It never was a cheap form of energy, in spite of early exaggerated claims, and the difficulties of storing radioactive waste are great. Thus, the technology of nuclear energy requires more technology to remedy the problems it creates. Or does it?

In France, over 70 percent of electrical energy comes from nuclear fission. How have the French managed to maintain a nuclear power industry without accidents and without overwhelming complaints from the population? The answer is a combination of technology, education, and politics. The technological fix makes use of a medium-sized nuclear reactor designed to operate reliably. The same design is used in all the power plants in France. Therefore, an engineer or operator trained to handle a plant in the north of France can also deal with an identical plant in the south of France. The training of technicians is extensive, using the stock of well-educated workers supplied by the French educational system. In the United States, it is difficult to find appropriate control-room operators because it is a tedious job where nothing happens for a long time, but when something happens the operator must know what to do instantly. Engineers find the job degrading; high school graduates are not educated enough.

It becomes apparent that the fix to this situation is not purely technical: it is educational and political as well. The American population must learn to accept the situation realistically and deal with it in a pragmatic manner. This has become quite difficult, for anti-nuclear extremists have made nuclear safety a topic that verges on pseudoscience. In the region surrounding Three Mile Island, an atmosphere of hysteria has led to an impression that large numbers of cancer deaths were caused by the infamous accident. Yet the latest epidemiological studies can find no excess cancer deaths attributable to the release of radiation from Three Mile Island.

In the field of medicine, improvements in technology have led to problems which are just becoming apparent. Yet they could have been predicted by anybody who sat down to think seriously about the situation. These problems arise, paradoxically, from the fact that the aim of medical research is to prolong life. In the early part of the century, this aim was accomplished by the conquest of infectious diseases such as diphtheria, scarlet fever, smallpox, polio, etc. Since these diseases primarily attacked the younger segment of the population, their eradication had the effect of greatly reducing the death rate among those less than twenty or so. The overall result was an increase in the life expectancy of the population from approximately 40 years in 1890, to about 75 years in 1990. This, everybody would agree, is a desirable consequence of medical research.

Another result is that those who live past 40 must now contend with heart disease, cancer, and other non-infectious diseases that develop gradually, and are seen mainly among the older members of the population. Medical research has shifted to alleviating these conditions— although it must be understood that in the case of cancer, a cure conventionally means that you live five years longer than you would have otherwise. Medicine does not defeat death. It merely postpones it.

One obvious effect of recent trends is that we see more and more centenarians enjoying life. Grandparents jitterbug and go white-water rafting. A ninety-year-old of my acquaintance recently passed away while vacationing in Acapulco.

That is the good news. The bad news is that we see more and more older people with senile dementia. Senile dementia is an umbrella term used to denote a number of conditions, most common of which is Alzheimer's disease, a physical deterioration of the brain. In 1987, there were over a million and a half severe cases of dementia in the

United States, a figure expected to double by the year 2020.[3] The basic reason for this increase is the rise in the number of older people. More than 80 percent of the people born in 1990 can expect to reach the age of 65, compared with 40 percent in 1900. Furthermore, the prevalence of severe dementia rises rapidly with age, as seen in the following table:

Age	Prevalence of Dementia
65–74	1%
75–84	7%
over 85	25%

There is no known cure for Alzheimer's disease. Even its cause is unknown. The technological fix for this disease is to find both the cause and a prevention. (Since the brain deteriorates so markedly during the course of the disease, a cure for severe cases is problematical.) In the meantime, the only recourse is to take care of old people who are gradually growing helpless.

A paradox is apparent: now that medical technology has extended the life of the population, we find that our success has doomed large numbers of the oldest age group to an existence in which they cannot remember who they are or who their children are, do not know what they are doing, and must be cared for at all times. The technological fix has yielded a bitter harvest. At one end of the process, we spend billions for research to provide longer life; at the other end, we spend billions to take care of those who lived too long.

Part of the solution to the problem is a human fix rather than a technological fix. As humans have greater life expectancy, they expect to live longer. In the United States especially, the demand for longer life has led to enormous expenditures on medical technology. Americans, with their faith in the technological fix, spend great sums of money on highly technical research and on highly sophisticated diagnostics and techniques. At the same time, the medical schools give a low priority to more economical measures such as public health and community medicine. A large fraction of medical costs is devoted to caring for patients during the last few months of life with the aid of sophisticated medical instruments and techniques. The same funds spent on immunization and nutrition for young children would go much farther toward improving public health and would be of greater benefit to greater numbers. The long-range consequence of this imbalance in attention is that medical costs are becoming an overwhelming factor in the national budget, and many students of economics are beginning to wonder if we can afford it.

One way to slow down this process is to cool the passion for longevity, to avoid procedures that cost great fortunes merely to add a brief period to an already long life span. Recently adopted rules require physicians to apply cardiac resuscitation to all patients who have not specifically forbidden this drastic procedure. The medical profession is recognizing that this rule produces terrible dilemmas and is painful to both patients and doctors. Patients with terminal cancer are violently brought back to life by chest pounding and electric shock simply because nobody in the family is willing to say "don't do it."[4]

The basic philosophy that has led us into these difficulties is a feeling among most people that death is the worst thing that can happen to a person. This is why the death penalty is considered worse than a lifetime spent in jail. However, a visit to a nursing home filled with demented elders can inject doubts into one's mind. Johann Sebastian Bach understood these matters well when he wrote *Komm süsser Tod*.

One solution to the problem of senile dementia is this: increase research into understanding the causes of and the prevention of senile dementia and decrease research on procedures that do nothing but prolong life without meaning or purpose. Until you understand dementia, don't magnify the problems. The fix is in part technological, but it cannot be accomplished without a change in human habits and philosophy. Can minds be changed rapidly enough to solve the economic problems before we are inundated with them? As an indicator of how difficult this is going to be, a recent paper in *Science* (estimating the maximum longevity that can be attained by curing disease) agrees with the above conclusion that more effort should be put into the prevention of dementia.[5] However, the authors choose not to take the risk of recommending that funds should be taken away from life prolongation research.

NOTES

1. D. H. Meadows, D. L. Meadows, J. Randers, W. W. Behrens III, *The Limits to Growth* (New York: Universe Books, 1972).

2. L. R. Brown, et al., *State of the World 1985: A Worldwatch Institute Report on Progress Toward a Sustainable Society* (New York: W. W. Norton & Co., 1985).

3. J. H. Gibbons, *Losing a Million Minds* (Washington, D.C.: Congress of the United States, Office of Technology Assessment, U.S. Government Printing Office, 1987).

4. E. Rosenthal, "New Rules for Saving the Dying Are Being Misused, Doctors Say," *New York Times* (4 October 1990): p. A1.

5. S. J. Olshansky, B. A. Carnes, and C. Cassel, "In Search of Methuselah: Estimating the Upper Limits to Human Longevity," *Science* 2 (November 1990): p. 634.

MYTH 15

Myths about Reductionism

Reductionism: n. any method or theory of reducing data, processes, or statements to seeming equivalents that are less complex or developed; usually a disparaging term.

The Many Faces of Reductionism

"Reductionism" is a term used by philosophers of science to designate the notion that complicated ideas or phenomena can be reduced to simpler elements. In physics, reductionism means the idea that everything is made of elementary particles and that everything that happens in nature results from the interactions of these particles. This concept seems obvious and inarguable to most physicists.

For example, I have shown earlier in this book that the behavior of fluids—their flows and waves and turbulences—can be reduced to the actions of the particles which make up this fluid. However, when we apply this logic to biology and approach the problem of explaining life or intelligence by elementary particle theory, then numerous objections may be—and are—raised. It is in this context that reductionism is so often used in a disparaging manner.

It appears odd that a method considered perfectly normal by a physicist can be maligned simply by calling it reductionist. Yet consider the following comment from a review of a book on complexity, nonlinearity, chaos, fractal geometry, and the like:

> Many, including this reviewer, believe the newly forming spectrum of ideas loosely called 'complexity' to be uniquely rich in its promise of deep insights that elude the standard reductionist methodology of physics.[1]

Apparently, according to the reviewer, the standard methodology of physics is reductionistic and lacks deep insight because it is based on linear logic and so avoids nonlinearity, feedback, and all the other intriguing features that typify complex systems. However, this comment is a straw man, for standard physics does contain all of those features that lead to chaos. Hydrodynamics is but one example. I will have more to say about this in the next section.

Book reviews are among the most common arenas for the use of "reductionism" as an epithet. See a *New York Times* review of *Sexual Personae: Art and Decadence from Nefertiti to Emily Dickinson:*

> "Sexual Personae" is tainted with the kind of symbol-mongering reductionism that sees one thing in everything, and despite its considerable virtues, it left me thinking of Earl Long's pithy appraisal of Henry Luce and his notoriously single-minded magazines: "Mr. Luce is like a man that owns a shoe store and buys all the shoes to fit himself. Then he expects other people to buy them."[2]

In this context, the epithet "reductionism" possesses the same flavor as the term "simplistic"—another favorite putdown—used to express the fact that the idea under criticism does not take into account all the infinite details of a more holistic view.

In order to understand the objections to reductionism, we must first detour to consider some of the varied meanings reductionism has acquired over the years. We will then see that most of the disputations arise from a confusion of meanings and that many disagreements can be banished by proper analysis. Other disagreements achieve an odor of theological contention and are not so easily clarified.

The controversy begins with a dispute between John Stuart Mill and William Whewell in the second quarter of the nineteenth century.[3] To Mill and his followers in the world of philosophy, the fundamental entities of the world are our individual sensations and observations. To them, science is the creation of theories by induction from particular bits of evidence. Causation is nothing but regularity in a succession of events; to them, how one event causes another is unknowable. They had no concept of microscopic causation as we know it now: the interactions between particles. "Reductionism," according to this school, is the reduction of all theories to the elementary *observables.* Only things observable to the senses are allowed to be part of a theory. The physicist Ernst Mach was a vocal supporter of this concept; theoretical constructs—even the notion of atoms and molecules—were to be avoided. These were not observables and so, at best, were merely useful hypotheses. As we saw under Myth 4, this attitude was a reaction against the mystical predilections of the phlogiston-caloric-ether school of physics.

William Whewell, leader of the antireductionist school, insisted that science was more than induction of generalities from particular cases, but involved the creation within the mind of organizing ideas and theories. To him, the hypothesis was as important as the facts, and a theory was to be valued for its explanatory and predictive powers. Whewell thought that knowledge was not derived entirely from the senses, but was a product of both sensations and ideas.

While Whewell's views appear strikingly modern to us, the ideas of Mills and Mach prevailed. Chemistry in the nineteenth century became so antispeculative that a prohibition on theoretical papers was established in the *Journal of the Chemical Society,* while the atomic hypothesis was dismissed as potentially misleading and metaphysical. Indeed, at a meeting of the Chemical Society in London in the mid-1870s, only one of the sixty members present declared himself to be a believer in the reality of chemical atoms. And this was seven decades after Dalton's initial proposal of the atomic theory and over a decade since Cannizzaro had used Avogadro's hypothesis to show how the atomic theory could explain a wealth of observations.

At the turn of the century, Pierre Duhem intensified the attitude of skepticism within the reductionist camp. In his view, the idea of atoms and molecules was simply a conceptual model used to represent what was going on within an observable chemical phenomenon. He considered the use of such models in science to be simply psychological conveniences (heuristic devices), and not logically necessary. To him, a scientific theory was a formal system of deductive reasoning expressing the relationships between observed phenomena. Such a system of deductive reasoning was considered better than a theory consisting of models and analogical reasoning.

In the early part of the twentieth century, the reductionist theories of Mach and Duhem led to the development of *logical positivism* by the Vienna circle. Logical positivism expressed the attitude that the only things you are allowed to talk about in science are observables. We see the spirit of logical positivism expressed in the work of the behaviorist psychologists, particularly B. F. Skinner, who carried the philosophy to such an extreme that he denied the existence of inner states of mind and even consciousness. All that counted was behavior, and psychology was to be reduced to the study of relationships among observed behaviors.

A peculiar turnabout occurred during the course of the twentieth century. Through the work of the physicists, atomic theory became so well established that it was no longer possible to attack it by calling it antireductionist. Philosophers had to reverse the position of reductionism in order to make atomic theory part of the establishment. Reductionism now embraced the idea that everything was made of atoms. As a result, the concept of reductionism now became the target of criticism from those who could not accept the idea that human beings were made of atoms and followed physical principles like everything in nature.

By the last quarter of the twentieth century, reductionism had evolved into a complex of ideas graced by a variety of definitions. *Methodological reductionism* is the application of research techniques borrowed from one scientific discipline to research problems of another discipline, e.g., using concepts of physics in economics, or using biological techniques in psychology research. *Theoretical reductionism* is the belief that the entire subject matter of one science can be presented in terms of another science, e.g., using neurophysiology to explain psychology.

The most basic kind of reductionism is called by some *ontological reductionism*. In philosophy, ontology is the branch of metaphysics dealing with the nature of being, reality, or ultimate substance. Ontological reductionism is nothing more than the basic premise of this book: the universe is composed of elementary particles and the forces by which they interact, *and nothing else.*

So confused is the situation now that the psychologist B. F. Skinner can be labeled as either a reductionist or an antireductionist, depending on which definition you like. He is placed in the reductionist ranks by the older definition (in which hypotheses about unobservable entities like the mind are not allowed). The newer definition (in which "the mind" is accepted as an abstraction and is considered to be the outward expression of nervous system activity) puts him in the anti-reductionist camp. Science would be in serious trouble if its technical terms frequently changed their meanings in such a violent and contradictory manner.

Working physicists—in particular experimental physicists—do not spend a great deal of time worrying about what philosophers say. They just do their work and let philosophers analyze and criticize. When 19th century philosophers said that atoms are simply models of no great utility, physicists blithely went ahead to invent instruments by which the mass, velocities, and electric charges of individual atoms and their nuclei could be measured. Then C. T. R. Wilson came along and invented the cloud chamber. Once an experimenter has seen the track of a particle passing through a cloud chamber, it is impossible for him to think that the object creating the track is anything other than a real particle. Within the past decade, a number of instruments—dubbed scanning probe microscopes—have been invented which directly image the atomic structure of solids, using feelers that pass over the surface of the material and measure tiny changes in electric fields or tiny variations in local atomic force.

The result of this instrumentation is that atoms are now accepted as real objects, with well-established properties. The habit of thinking of particles as the real building blocks of matter is now firmly ingrained in the psyche of physicists. Using the equations of theoretical physics to predict how these particles are going to behave is the only way that physicists know how to think. When critics and philosophers discuss how physicists think, they call it reductionism. But physicists hardly ever stop to think that they are engaging in reductionism. (It's like the person in Molière's joke who doesn't know he is speaking prose.)

The turnabout in perspective during this century has been astonishing. Originally the reductionists refused to admit the very existence of atoms. Their elementary entities were human sensations. Now the reductionists insist that atoms are real things and the elementary entities on which their theories are built are particles far below the level of atoms. The new reductionists are pragmatists. Experience has taught them to trust their instruments more than their raw sensations, and the instruments have opened up the microscopic world which underlies reality.

It is clear that many of the disagreements concerning reductionism are due to the fact that—in different periods of time and in different disciplines—there have been a variety of basic elements to which scholars habitually reduced their theories. Previously, the basic elements were sensations—the input from sensory organs. Now they are the elements of nature observable by our instruments: particles, atoms, molecules, and the cells of the nervous system. Is that all there is? On this question the battle is joined.

Linearity vs. Holism

Jonathan Kellerman is a mystery writer whose detective character happens to be a clinical psychologist. In one of his stories, the psychologist is explaining to a young client the data analysis involved in a psychological study.[4] The client says, "Kind of reductionistic, don't you think?"

The psychologist asks, "In what way?"

The client responds, "You know—testing us all the time, reducing us to numbers, and pretending the numbers tell the truth."

Thus, reductionism has invaded popular literature, and as is so often the case, it takes the form of an intellectual putdown. Psychology, as an inexact science, is most vulnerable to charges of reductionism. How can psychologists operate without oversimplifying?

Frequently, when a critic uses the term reductionism, it appears to be a subtle mask for a set of theological preconceptions. Archaic concepts such as *elan vital,* psychic energy, and soul may generate resonances in his memory. When a psychologist or a neurophysiologist tries to argue that consciousness and emotions can be explained as nothing more than the action of a physical nervous system, the antireductionist may then say "but human beings are more than computers, more than just heaps of atoms and molecules." The scientist must then ask: what does the "more" consist of? Is it a new principle of science or is it something of a supernatural nature?

This question leads a member of the Department of Communications and Neuroscience at the University of Keele in Staffordshire, England, to assert the following:

Reality can have many levels, all of which must be reckoned with in appropriate circumstances if we are to be truly realistic. This point seems to be missed in much popular scientific writing about human beings. Instead there is a subtle (or sometimes blatant) indoctrination of the readers with the fallacy of ontological reductionism or 'nothing-buttery.' . . . To disparage other accounts of human nature in the name of science, on the grounds that humans are really just a network of biological computing systems, or really just naked apes, or really 'throwaway survival machines for our immortal genes' (blurb on the cover of Richard Dawkins' *The Selfish Gene*) is, wittingly or otherwise, to attempt to deceive the public. It is simply false to suggest on any of these grounds that we are warranted to deny or neglect more traditional assessments of human beings as conscious moral and spiritual agents.[5]

The statement starts with a call for realism, yet by the end of the paragraph the evocation of moral and spiritual agents gives a distinct flavor of theology to the dispute and makes one wonder if there is not a hidden agenda: an effort (conscious or unconscious) to combine science and religion.

The same arguments take place in biology. Is the difference between living matter and nonliving matter simply a matter of chemistry? The biologists have a simpler time of it than the psychologists. At least the biologists feel that with the current understanding of DNA and gene structure, we are very close to understanding the mechanism of living organisms. Indeed, as I write, there is an announcement by Julius Rebek, Jr., a chemist at the Massachusetts Institute of Technology, who has created a compound (amino adenosine triacid ester) which has the ability to reproduce itself.[6] It does this by grabbing two simpler molecules, holding them in place until they combine, and then releasing the newly assembled molecule which is a duplicate of itself. In this way, the AATE acts as a templet. Other chemists are working on molecules which have the ability to assemble themselves from simpler components without using a templet. This self-replication property is believed to be necessary for living matter to evolve.

The chemists trying to synthesize living matter are working from the bottom up; they assume that the atoms of living matter are the same as the atoms of nonliving matter. The only difference to them between living and nonliving matter is the way the atoms are organized.

Many antireductionists try to explain that there is a fundamental difference between the "linear" laws of the physicists and the "nonlinear" or "holistic" thinking of those trying to deal with the complexities of living beings. This difference appears to have become a major issue in New Age philosophy.

See, for example, an interview of astronaut Edgar Mitchell in *New Frontier.*[7] The interviewer asks, in reference to the new perspective on life acquired by Mitchell while orbiting the moon: "Did this new perspective shatter the typical scientific reductionist conviction, and give you a holistic viewpoint of life?"

Mitchell replies: "Not necessarily, it merely helped me to see things in a more balanced perspective. Reductionism has its use. It's a very powerful tool; however, it's not the only tool. I now recognize the universe has its holistic point of view, that it is interconnected, that everything is a part of everything else. Energy and information flow from the microscopic to the macroscopic, back and forth."

Mitchell avoids the effort by his New Age interviewers to form a dichotomy between reductionism and holism. He understands the feedback between the microscopic and the macroscopic. When your finger approaches a flame, the microscopic nerve endings send signals up to the spinal cord and the brain. The sensation of heat is a macroscopic phenomenon involving large areas of the brain. Very quickly feedback goes back to the lower levels; impulses travel along the nerves to the muscle fibers, causing the finger to withdraw.

The physics of heat and the response of molecules to heat represent linear phenomena. The operation of the entire control system, complete with feedback, is a holistic phenomenon.

Precisely what do we mean by "linear" and "holistic"? Linear in physics can mean a variety of things, making it difficult to understand what the New Wave people mean. In mechanics, two forces are linear if the forces add together like numbers: one force unit plus one force unit equals two force units. In electronics, a linear amplifier is an amplifier whose output signal looks just like the input signal, except that it might be larger or smaller. In complex systems, linear behavior is a simple kind of behavior—for example, a spring that obeys Hooke's law. In a nonlinear spring, the stretch is not proportional to the applied force. Human behavior is nonlinear because stored programs and feedback loops cause the response to a stimulus to be somewhat unpredictable. A simple linear chain of causality has no feedback loops.

Linear thinking, in this context, seems to mean what I have previously referred to as bottom-up thinking: looking at events from the viewpoint of elementary interactions between elementary particles.

Holism, on the other hand, means looking at a complex system as a whole, from the top down, taking into account all the interactions between the parts of the system, taking into account all the feedback loops and the effects of learning and reasoning and emotion.

What I want to show is that there is no dichotomy between linearity and holism. They are simply two ways of looking at the same thing. Furthermore, scientists—in particular information scientists and systems theorists—are well aware of the concepts and methods of holistic thinking. To say that scientists do not understand these matters is to perpetrate a myth. Thinking in this area goes back to the 1940s and 50s, when the mathematicians Norbert Wiener and John von Neumann wrote of analogies between human beings and computers.[8]

A holistic system possesses one thing that particles do not. That is *information.* Information is contained in the way particles are arranged: in structures, clumps, waves, pulses. Information is either stationary, stored like the data on a computer disk, or else it is transmitted through space in the form of modulations of some kind of energy.

In information science, "information" is purely a mathematical concept. Information theory deals with matters such as the amount of information (i.e., the number of signals) that can be transmitted through a given channel in a given amount of time. It says nothing about the "meaning" of the information. The top level, the place where biology and psychology coalesce, is the region in which we try to understand how the signal pulses acquire meaning, how the patterns of light and dark become a picture, and how we become aware of the pictures and meanings. Scientists have not yet reached an understanding of that level.

Yet we can see dimly how these questions might be answered. In a computer, information consists of a series of electrical pulses. These pulses, in turn, consist of groups of electrons passing through a circuit. In a living organism, information may be resident in the patterns of nucleotide base pairs in a DNA molecule, or in the patterns of amino acids in a protein molecule, or in the pulses of electrons and ions passing through the synapses of the nervous system. In any kind of mechanism or organism, information is built of electrons, atoms, and molecules, just as a house is built of boards or bricks. A house is more than a pile of bricks; it is bricks arranged in a certain way. But still—

from a material point of view—there is nothing in it other than its boards and bricks. The only thing extra is the information contained in the pattern made by the boards and bricks.

Only when we look at information from the top down do we recognize it as information. From this holistic point of view we can see a progression of electrical pulses making its way through a circuit instead of just an incoherent rush of individual electrons. But where does meaning emerge from the information? We can get a glimmering of an idea by developing an electronic device from the bottom up.

In an electrical conductor, such as a metal, the outer-most electron of each atom is not connected tightly to that atom, but floats in a "sea" of free electrons within the crystal lattice. If an electric field is applied to the conductor from some outside source such as a battery, then the electrons will move easily from the negative to the positive pole of the battery. We call this flow of electrons an electric current.

In a semiconductor such as silicon, there are only a few electrons free to travel among the atoms they were originally attached to. Adding minute quantities of certain impurity elements to the silicon (doping the crystal) increases the number of free or "conduction" electrons. The result is an n-type semiconductor in which the electric current is carried by negatively charged electrons. Another type of impurity added to silicon produces a semiconductor in which there is a deficiency of electrons, so it behaves as though the electric current is carried by positively charged "holes." We call this a p-type semiconductor.

If an alternating voltage—a sine wave—is applied from one end of a piece of semiconductor to the other, the current that flows also has the form of a sine wave. The semiconductor responds to the voltage in a *linear* manner; the output looks just like the input.

But an astonishing thing happens if an n-type semiconductor is closely joined to a p-type semiconductor, forming a device called a diode. The boundary between the two parts of the diode is called an n-p junction, and it is this junction that causes all the marvelous properties of the diode to emerge. What happens is that electrons and protons line up on opposite sides of the junction like football players on both sides of a scrimmage line. The p-side becomes negatively charged and the n-side becomes positively charged. The resulting electric field acts as a barrier to electrons trying to flow from n-side to p-side.

Now comes the crucial part of the story. If you connect the positive end of a battery to the n-side of the diode, the potential barrier is increased, and, as a consequence, the diode acts as a high resistance. Reversing the battery makes the diode act as a low resistance. The diode behaves differently, depending on how the battery is connected. This means that when an alternating voltage is applied to the diode, you do not get an alternating current flowing through it. Instead, you get a series of current pulses, all going in the same direction. The output of the system no longer looks like the input. *We have created a nonlinear system by putting together elements which individually are linear.*

The fascinating thing is that all through this process, each electron simply does what it has to do, responding in a linear manner to the applied electric field as well as to the electric fields within the solid. But because of the ability of positive and negative charges to separate from each other, an additional internal electric field is created across the junction. This electric field essentially adds feedback to the system so that its total response is nonlinear.

I dwell on this point because of critics (such as the authors of reference 1 and 7) who complain that physicists and chemists look only at the linear interactions between the microscopic parts of a system and ignore the nonlinear effects that emerge when the microscopic parts are combined into a larger structure. But this emergence of new properties from simple fundamental processes is the very basis of solid state physics (as well as many other areas of physics). Thus the criticism is unfounded.

Even more interesting effects emerge when we combine n-p junctions into more complex arrangements. Join two diodes together and you make a transistor which has two junctions within it and three connections to the outside world. A signal applied to one portion can control the flow of an electric current passing between the other two parts. The transistor can act either as an amplifier or as an on-off valve. In this latter capacity it plays the leading role in digital circuits, in which the absence of a current is interpreted as the number zero and the presence of a current is interpreted as the number one. We begin to see information emerging from the operation of this system.

Connecting two or more transistors in an appropriate manner creates the various circuits used as components of digital computers: memory circuits, arithmetic circuits, and logic circuits. But in all of this complexity, each electron and proton does no more than what it has to do. Each electron finds itself in an electric field and moves in the direction of that field. That is all it is able to do. We do not notice that the system is processing information until we get above it and look down to see that the circuits are manipulating pulses of electric current—each pulse being nothing but a horde of electrons moving through a wire. We start to think that the notion of information is something invented by humans to describe something that is observed by human observers. As for the "meaning" assigned to the information, surely that is something that only humans can create. However, I think that it is possible to define "meaning" in such a way that it does not always require a human interpreter.

Let us use as an example a device that by now is familiar to most inhabitants of the industrialized world: the remote control used to operate a TV receiver. The remote control is merely a generator of electrical pulses which are converted by a light-emitting diode into pulses of infrared photons that are absorbed by a sensing device (a photocell) in the TV receiver. The pulses are organized into a set of digital codes, somewhat like the code bars read by a supermarket checkout computer. One code makes the audio louder, another code switches channels, another code causes the TV to display the time, etc.

I propose that the response of the TV receiver represents the "meaning" of the information transmitted by the remote control. This is an operational definition as well as a behavioral definition. If the code makes the sound volume increase, the meaning of the code is "volume increase." If the code makes the channel increase by one, the meaning of the code is "channel up one." This definition is undoubtedly inadequate to cover all the meanings of meaning. But it serves nicely as a partial definition whose purpose is to demonstrate that when information is received by a nonhuman device of some kind, this information is given meaning by the action taken in response to the message. It is a definition a behaviorist psychologist would understand, but I think it is not a complete definition.

Note that the information takes on meaning by going from one device to generate an action in another device. Information can also travel from one part of a device to another, as when you talk to yourself. The part of your brain that understands language doesn't care whether the signals come from the ear through the auditory nerve, or whether they are generated in another part of the same brain.

In taking you through the development of a digital circuit from the interior of a diode to the upper levels of a computer system, I have tried to show how at each level a different mode of description applies. Within the diode, quantum theory and solid-state physics works well. When we connect diodes into transistors, higher-level descriptions called circuit equations are more convenient. When we connect transistors into complex circuits, the circuit equations may or may not work; if not, systems theory must be applied. At all levels, what happens is nothing but the motion of electrons and photons; an observer sitting on an electron would be aware of nothing but the electron moving along electric field lines. Our appeal to multiple levels of description is for purposes of convenience so that we may understand and describe what happens without following the motion of every individual particle. Understanding the action of the circuit requires both reductionism and systems theory.

Linearity and holism are both parts of the same truth.

The Hidden Agenda

What can we do about the confusion surrounding reductionism? One thing we can do is to ignore it. Scientists in their workaday roles never ask themselves whether they are engaging in reductionism. They just go ahead and do their work. A particle physicist does not feel guilty about indulging in ontological reductionism when he assumes that everything in the universe is made out of elementary particles. That's just the way he looks at the world. A neurophysiologist who tries to determine how the nervous system provides us with a picture of the outside world assumes that his task is possible without inserting mystical concepts into his model. Otherwise, his goal could never be reached and he would not be able to do his work.

The people who take reductionism seriously are book critics, philosophers, and New Age enthusiasts. Many of these are honest analysts who assume the task of keeping scientists honest. However, when I read what some critics of reductionism have to say—particularly in book reviews—I suspect that a hidden agenda lies beneath their antireductionist attitude.

What sort of hidden agenda do I mean? Before I answer that question, let me first reexamine the most elementary definition of reductionism: the idea that everything in the universe is composed of elementary particles interacting according to known laws—*and nothing else*. What could be simpler? In this decade, a century after the discovery of the electron, what could be less contentious? To some individuals, however, the "nothing else" clause raises a great deal of concern. (Recall the complaints about "nothing-buttery" science in the previous section.)

Is it true that we can account for intelligence, feelings of love, hate, joy, and spirituality, along with all other human attributes, by assuming that we are all "nothing but" a collection of particles? In trying to answer this question, it helps to focus matters if we turn it around and ask: if there is something in a human being above and beyond the elementary particles and the elementary forces, what is that extra ingredient?

There are three possible kinds of answers—three kinds of extra ingredients.

1. One answer is that there is no additional material, theoretical, or spiritual ingredient. The known particles make up all there is, and the higher-level activities we observe in living things are enabled solely by the way the particles are organized. The organization of living matter is the qualitatively different feature, the ingredient present in addition to "just particles." This is what we discussed in the last section: the ability of molecules to form self-replicating forms, the ability of certain kinds of circuits to manipulate information and extract meanings out of that information—all the processes we lump under systems theory. If this kind of complexity is enough to explain everything humans do, then we need go no further. Nothing *physical* is needed to complete the explanation. Within this reductionist model, nothing exists anywhere in this ladder of complexity except elementary particles doing what they have to do according to the laws of nature.

This is the position taken by Erwin Schroedinger in his little book *What is Life?*[9] He asks the question "Is Life based on the laws of physics?" and answers: "From all we have learned about the structure of living matter, we must be prepared to find it working in a manner that cannot be reduced to the ordinary laws of physics. And that not on the ground that there is any 'new force' or what not, directing the behaviors of the single atoms within a living organism, but because the construction is different from anything we have yet tested in the physical laboratory." This was written at a time when information theory was in its infancy and the concept of feedback in machines (cybernetics) was just being explored.

2. Some people are unsatisfied with this answer. The possibility that systems theory can provide all the answers is unsatisfactory to them. They propose that something new and radically different must be added to science as it is presently understood to breathe life and intelligence into humans. Some nod in obeisance in the direction of "quantum reality," hoping that will be the new reincarnation of *elan vital.* Others, notably Roger Penrose, propose adding novel physical mechanisms such as "correct quantum gravity" to physics as it is presently understood (See chap. 7, ref. 8). Neither of these approaches provides a mechanism for answering questions about thinking, feeling, and consciousness. My opinion is that it is premature to start evoking new entities. We have not yet proved that the present entities are insufficient to solve the problem. Neurophysiology and computer science are such new fields of study that we are not close to exhausting the possibility of explanation by natural means, using the particles and forces presently known.

3. Members of the third group believe that living matter is truly different from nonliving matter and that explanations of consciousness require the introduction of mystical entities such as psychic energy. This group clearly operates from a religious background. Their beliefs make it hard for them to think that human beings can have human characteristics without the intervention of supernatural forces. The difficulty with their argument is that they start out with nothing but a belief that science is insufficient to explain life. They do not have any specific mechanisms capable of supplying understandable explanations. They complain that reductionist theories are simplistic, but have nothing of a less simplistic nature to replace them with.

The concept of supernatural forces presents a paradox to science. If supernatural forces were real in the sense that they could be identified, measured, and their laws determined, then they would become part of nature. They would no longer be supernatural. On the other hand, if supernatural forces work completely outside the domain of the natural, then how can they have physical effects on things within nature? How do these forces make connection to the corporeal body? Where do these supernatural forces originate? One can believe in ce-

lestial voices only if one has an idealistic view of the mind and believes that the mind is an entity unto itself, is distinct from the body,
and possesses a mystical connection to a higher realm.

Which brings us back to the topic of hidden agendas. The agenda,
in the case of group 3, is at least out in the open with its insistence
on mystical entities. The members of this group unquestionably are
motivated by faith. They do not want to admit that living beings can
operate without a prime mover.

The second group is the most problematical. Its members try hard
to assume a position of scientific rigor. Yet, in their impatience with
realistic science and in their attempts to read mystical meanings into
quantum theory, it appears to me that secret motivations underlie their
assumptions. They oppose "nothing-buttery" science because the idea
that everything is made of nothing but elementary particles rules out
the possibility of the supernatural. The secret agenda in their minds
is to reverse the secularization of science that has been going on for
the past century and introduce philosophical ideas that originate in
religion.

Occasionally a critic of reductionism gives the game away by coming completely out into the open. Consider a review of an earlier book
published in what is obviously a newspaper with a fundamentalist slant.
While the review was actually quite fair as far as subject matter was
concerned, one comment was significant:

> Milton Rothman is an atheist. I'm a Christian. What do I think of this
> book of his? Well, even though he's trying to teach atheism and not just
> science, Chapter 1 does give an interesting account of how physics and
> similar sciences work. . . . In his last three chapters Rothman goes on from
> physics to atheism ('reductionism').[10]

The parenthetical remark indicates that in the reviewer's mind there
is a clear equation of reductionism with atheism. And here we get to
the core of the matter. If reductionism equals atheism, is the converse
true? Is antireductionism a screen for religion? In the case of the review
quoted above, there is no doubt about it. I think that in many other
cases, criticisms of reductionism come from writers attempting to be
scientific while a religious soul is pushing out from behind the curtain.
And that is the hidden agenda.

Negative Reductionism

Why is reductionism a dirty word? It is because most people find it impossible to explain the workings of the human mind starting with the elementary particles, the fundamental laws of nature, and nothing else. For this reason, they believe additional ingredients are required. The argument is only a straw man, for no scientist in his right mind would ever attempt to deduce the structure of the brain working bottom -up from particle theory. Scientists well understand the difference between particle theory and systems theory, and know how to progress from one to the other.

The attempt to bring in supernatural entities is fallacious for another reason. The kind of logic used goes something like this:

1. Nobody knows how to figure out the working of the human mind.

2. Therefore, it is impossible to understand the working of the human mind on a natural basis.

3. Therefore, it is necessary to bring in supernatural entities to explain how the mind works.

The second statement does not follow from the first. Not knowing how to solve a problem does not make the solution impossible. At the same time, we do not know yet whether the solution is possible.

Fortunately, there is another way of attacking the problem. Let us, for the moment, forget about explaining life. Let us not try to explain how the brain works (not in this book, at least). Trying to explain and predict what human beings do is too difficult for us. Let us, instead, predict what people *cannot* do. This is what we have been emphasizing all along in this book, for negative predictions are based on the fundamental symmetries of nature—and so apply to all the particles of the universe, no matter how they are arranged or organized. Since they apply to the individual particles, they also apply to the objects built up out of these particles. (We have already discussed how this happens in chapters 1, 3, and 13.) Predicting what *cannot* happen makes negative reductionism a powerful tool.

Here is a review of some of the predictions we have made in this book:

• Nobody is ever going to build a perpetual motion machine (or equivalent device). (Conservation of energy)

• Nobody is ever going to hang levitated between floor and ceiling without material support or other physical force such as magnetic fields. (Conservation of momentum)

• Nobody is going to travel to the distant stars at speeds faster than that of light. (Principle of relativity; Poincaré symmetry)

• Nobody is going to send messages of any kind at speeds faster than that of light. (Poincaré symmetry)

• Nobody is going to send any kind of message that does not get weaker as it travels away from the source. (Heisenberg uncertainty and conservation of energy)

• Nobody is going to send any message through space without the transmission of energy by a physical carrier (particle or photon). (Conservation of energy)

• Nobody is going to send messages directly from one mind to another. (At least not without an amplifier. Electromagnetic fields produced by the brain are too small to carry signals over any appreciable distance, and no other signal carrier is known.)

• Nobody is going to receive messages directly into his/her mind without the agency of a physical carrier. (This includes perception at a distance as well as perception of future events.)

All of these negative prophecies follow from the assumption that everything is made of fundamental particles that obey the laws of nature, and that there is nothing else. The structure of the human brain, the nature of consciousness, or the meaning of life are irrelevant to this kind of logic. These are specific predictions that can be made from the position of ontological reductionism.

If you want to demonstrate that reductionism in this sense is incorrect, you must show that one of the above statements is wrong. These are specific tests for the validity of the science that we now know. The past century has been a continuous testing ground for these and similar principles. Relativity, the nature of the fundamental forces, the symmetry principles, and communications theory have been under the most intense scrutiny. Demonstration that one of the statements above is false would be the most unexpected and important development in science.

But as we survey what has happened during the past century, we find that the validity of our predictions becomes more certain as time goes on. Parapsychology supporters are still tilting at their windmills, but realistic scientists are unworried.

What is most important: a survey of the history of science for the past three centuries finds an important new principle emerging. This is a theme that threads its way throughout this book. What we are beginning to recognize is that throughout history, every scientific discovery of a permanent nature has been based on the philosophy of realism. Every theory with an idealistic or unrealistic flavor has fallen by the wayside.

This means that anyone embarking on parapsychology research is taking an enormous risk. It is, of course, not possible for me to extrapolate. I cannot say that because realism has been right in the past, it will be the only possible way of doing things in the future. However, an appreciation of history and the direction of current discovery makes it easy for one to see that molecular biologists, neurophysiologists, and information scientists are increasing their knowledge at an overwhelming rate, while parapsychologists are floundering about, getting nowhere very fast. It should be easy for a young scientist to decide where to put his time and effort.

NOTES

1. P. Carruthers, in a review of *Exploring: an Introduction,* by G. Nicolis and I. Prigogine, in *Physics Today* (October 1990): p. 96.

2. T. Teachout, review of *Sexual Personae,* by C. Paglia, in *New York Times Book Review* (22 July 1990).

3. R. Harre, "Philosophy of Science, History of," in *Encyclopedia of Philosophy,* ed. P. Edwards (New York: The Macmillan Co. & The Free Press, 1967), vol. 6.

4. J. Kellerman, *Over the Edge,* (New York: Signet Books), p. 25.

5. D. M. MacKay, "Seeing the Wood and the Trees," in *Thinking: the Expanding Frontier,* ed. W. Maxwell (Philadelphia: The Franklin Institute Press, 1983), p. 8.

6. M. W. Browne, "Chemists Make Molecule With Hint of Life," *New York Times* (30 October 1990): p. C1.

7. Sw. Virato and Djuna Wojton, "Edgar Mitchell—New Age Astronaut," *New Frontier* (January/February 1991): p. 13.

8. N. Wiener, *Cybernetics, or Control and Communication in the Animal and the Machine* (New York: John Wiley & Sons, 1948); N. Wiener, *The Human Use of Human Beings* (New York: Houghton Mifflin Co., 1950); J. von Neumann, *The Computer and the Brain* (New Haven: Yale University Press, 1958).

9. E. Schroedinger, *What is Life?* (New York: The Macmillan Company, 1945), p. 76.

10. A. Lohr, review of *A Physicist's Guide to Skepticism,* by M. A. Rothman, in *Chattanooga News-Free Press* (24 July 1988).

MYTH 16

"Myths are just harmless fun and good for the soul."

Good Myths

The popularity of myths is evidence that they serve a useful function. It is probable that humans could not exist without some form of mythology, for every person born with curiosity retains a deeply imbedded need to explain in some manner the world within and without. Prescientific humans (which includes most living humans) use myths as a means of providing simple explanations for the origin and workings of the universe. It is much easier to believe that humans appeared full-blown in the Garden of Eden than to understand the complex mechanisms of evolution and molecular biology. It is easier to believe that "fate" brought you to your present predicament than to analyze the complex web of decisions made by yourself and by all the people around you. It is easier to say, "The devil made me do it," than to take responsibility for your own actions.

Furthermore, as many have pointed out, myths, fantasy, and fairy tales have a function in the development of children.[1] Hero myths provide motivation for slaying dragons, overcoming evil, and surmounting obstacles. Cautionary myths provide instruction in ethics: good people live happily ever after. Mystery myths, as exemplified by Sherlock Holmes, teach observation, deduction, and logical thinking. Fairy tales teach racism, both positive and negative. The frog might be a prince, but the dwarf, or troll, or anybody else who looks different is undeniably bad.

From all this arises a *metamyth*—a myth about myths: *myths are good for you and are also great fun.* There is no denying that the world would be less pleasurable without myth and fantasy. We might go so far as to say that for most of the people most of the time, myth and fantasy make up *all* the fun in life. Without myths the world would be a drab place.

On the other hand: in the political world of the United States, the public is discovering (even as I write this) that stripping the shield of myth away from a government displays the nation in a state of confusion and despair. For over a decade, the elected officials in Washington propagated the myth that it was good for the government to receive less in the way of taxes and to spend less on services it claimed were unnecessary. This myth was received with great joy by the public which, reveling in lowered tax rates, elected the same party three times in a row. The reality, however, was that while taxes decreased, the government did not reduce spending proportionately, with the result that the decade of the 1990s found governments at all levels head over heels in debt, and with insufficient money to operate legally. In addition, the myth that "the less government regulation the better" spawned the financial bubble of the 1980s, with excesses of risk-taking and corporate takeovers. It was great fun while it lasted, but then the bubble burst, as bubbles tend to do.

The previous example exposes the dark underbelly of mythology. Like so many other enjoyable pastimes such as drinking, drug using, and sex, their indiscriminate and uncontrolled use may have deplorable consequences. Most unfortunate is the fact that acting on the perceived injunctions of mythology may have worldly effects that are so slow in becoming evident that by the time we notice something is wrong, it may be too late to fix it. Such is the case with the myth, discussed in the last chapter, that population growth is good for the world. Other myths produce their damage more quickly, and those we want to examine now.

Bad Myths

A myth is bad when its consequences produce more harm than good. Since a myth invariably disguises reality, any myth that camouflages natural dangers leaves the believer unwary of the hazards existing in the world.

This is not just paranoia. During the period of sexual freedom that opened up during the 1960s, the myth arose that unlimited sexual expression was the ultimate human freedom. In the post-Freudian world, sexual repression was advertised as the cause of all mental disorders. (What Freud actually said was that *too much* sexual repression causes *some* mental disorders—notably hysteria. Whereupon those who wanted to eliminate all sexual repression expanded Freud's statement into a myth.) Therefore, since we knew how to handle fertility, there was no logical reason to forbid any form of sexual behavior. "Tune in and turn on" was the motto.

Unfortunately, the postwar generation forgot what all those World War II training films had pointed out so clearly to my generation: in nature there exist sexually transmitted diseases, and if you indulge in indiscriminate sexual intercourse, you have a good chance of acquiring one of those diseases. Then along came AIDS, and the public abruptly became aware that with sexual freedom come sexual dangers. Most of those in the homosexual population—the population at greatest risk —suddenly became quite circumspect in their sexual contacts. Reality dictated prudent behavior. It was not necessary to be intimidated by "moral standards" imposed from above, either by the law or by theologians. Prudent behavior arose from a knowledge of reality and from an intelligent desire to stay healthy.

A certain portion of the population subscribes to the myth that thinking the right thoughts is a protection against disease. It is true that for certain types of conditions, particularly disorders of psychological origin, positive thinking can make you feel better. It may even be that the brain can control the workings of the immune system sufficiently to fight against certain infectious diseases. But there is no evidence that psychology can work miracles against strictly physical disorders such as cancer, broken bones, intestinal obstructions, etc.

James Randi has documented the fraudulent claims made by faith healers and the credulity of the public toward those same healers.[2] It is possible to make the argument that for most of these people, their belief in faith healing is a harmless preoccupation. So what if they put up some money when they go to a faith healing meeting? The cost can be charged to entertainment. If it makes them feel good, why is it worse than spending the same amount going to a rock concert or an opera?

But for some people, of course, the entertainment is not harmless. If, for example, an insulin-dependent diabetic goes to a faith healer rather than to a physician, he or she is sure to end up in a coffin before the year is out. From the libertarian point of view, each person has the right to leave this mortal coil in any manner desired. This is all well and good. But the person taking the risk may not really want to die immediately, and the cost to the families and others left behind can be considerable.

The great tragedies occur when subscribers to the myth of faith healing subject their own children to their irrational beliefs. In a recent case that received wide publicity, a Boston couple, members of the Christian Science church, tried to treat their two-year-old son with prayer. The boy was suffering from an intestinal obstruction, a physical condition readily diagnosable and treatable with surgery. The parents did not even look for a medical diagnosis, the boy died, and the parents were convicted of manslaughter.[3] Here was a case where the death of a child was directly caused by the myths of the parents. The myth in this case was not harmless fun.

As might be expected, conviction by jury failed to eliminate the delusion. The father was quoted as saying: "The verdict will strengthen the conviction of Christian Scientists to do stronger and better spiritual healing. We have to strive for a perfect record." The wishful thinking expressed here is appalling, for this was the fourth case in 15 months in which Christian Scientists were convicted of either manslaughter or neglect in the death of their children.

A response often made to this story is, "But if the boy was operated on, he might have died anyway. Doctors have been known to lose patients, also." Comments like this are evidence of mathematical illiteracy (innumeracy), and an inability to think in a statistical way. A more rational way to deal with this situation is to compare the risk of dying from faith healing with the risk of dying from surgery. So we ask: what is the probability of dying from complications of surgery? This is not an easy figure to come by. The best I can find is listed under a category called "Misadventures to patients and abnormal reactions or complications." This can include all kinds of accidents taking place in a hospital, such as abnormal reactions of patients to medication. (In other words, it is not always a human error.) In Philadelphia in 1982, there were 37 deaths of this nature out of 20,051 total deaths.[4] Translated into statistical terms, this figure means that the probability of dying from medical accidents of all sorts is about two chances out of a thousand. Not totally negligible, but not overwhelming.

In order to really understand this number, you must compare the chance of dying from one cause with the chance of dying from another cause. There are times when doing nothing is more dangerous than doing something. This is especially true in the case just cited, where the probability of dying from the untreated intestinal obstruction was one hundred percent. In other words, there was total certainty of death from the faith healing procedure as compared with less than two chances out of a thousand if the patient had been seen by a good surgeon. What kind of gamble would you take?

One response to this on the part of the believers is that it is more important for the parent to retain his faith than to gamble on medical science, in which he has no faith at all. The faith is important for the parent's peace of mind. The court disagreed. Libertarians might argue that a person has the right to endanger his own life through faith healing. But when this principle is forced upon a child unable to understand the implications, the result is manslaughter.

We begin to see that there is another myth at work here: the myth that doctors don't know anything for sure, a variation of Myth 2 (nothing is known for sure). While it is true that the assuredness of medical knowledge rarely approaches the precision of knowledge in physics, it is a gross exaggeration to claim that physicians know nothing. In this case, the mythologizer chooses the faith healer who claims true knowledge, but who really knows *nothing* for sure, in preference to the doctor who knows many things fairly well. It's not a good gamble.

Some myths have nothing to do with the supernatural, but are insidious in that they sound perfectly reasonable and dangerous in the sense that they have harmful effects on individuals. One is the myth that "all humans have equal capabilities." This is essentially a political myth, since it stems from the statement in the Preamble to the Declaration of Independence that "all men are created equal." Even as a political statement this was a myth from the start, since Section 2.3 of the Constitution defined each black to be worth three-fifths of a free person, while Native Americans were worth nothing at all.

Modern politicians interpret the equality clause to mean that all persons are created with equal political rights: the right to vote, the right to an education, the right to get a job, etc. The implication is that once a person gets into school, you can push him through the mill the same as everybody else. From then on the adults are on their own. Nobody guarantees that the outcome is going to be the same for all people. If you don't do as well as somebody else, it's your own fault.

What makes this myth so dangerous is that the reality of contemporary schools guarantees that the inequality with which human beings are born is going to be amplified by the schools and become as great as possible by the time they finish. It is clear that some people are born with advantages lacking in others. Some people are born with entrances into the best private schools and with a built-in social network. Others are born with no entree into the mainstream of society and no hope of attending superior schools. They are born instead with both hardware and software damaged by the environment—lead poisoning, fetal alcoholism, drugs, lack of parents, bad schools, poverty, the isolation of ghetto life, etc.

Let me make an important disclaimer here: I am not claiming any kind of inferiority based on genetic transmission of racial characteristics. At this point, I am interested in differences between individuals, not on differences between groups. After we get rid of environmental handicaps and level out the playing field, we can worry about genetics.

What are the consequences of the myth of equality? One consequence is that when we treat all students equally and try to push them all ahead at the same rate, we are effectively handicapping the slow learners, for the slow ones will become discouraged at their lack of progress and drop out. If we want to reduce the handicap, we must use an educational system that helps the slower learner, while at the same time liberating the faster students to go as fast as they can.

This system will not result in a society in which all citizens are equal in ability. The outcome—even under the best of conditions—will be a population in which some are faster than others and some are smarter than others. But the differences in the two groups will be less than they are at present because we will lose fewer of the slower ones.

Unfortunately, even under these ideal conditions, the conditions set by society adds a further blow to the lives of those at the margin. Smart people have a habit of increasing the level of technology used in industry so that it requires smarter people to work the machines. Managers brag about the new labor-saving devices that replace workers. Where those workers are to go is not their problem. The result is fewer jobs available to those with lots of muscles but less intellectual capacity (fewer farm jobs, fewer lumbering, ditch-digging, and laboring jobs). The result is what formerly was called "technological unemployment," and then— more euphemistically—"structural unemployment." The change from "technological" to "structural" is a hint that there is something discomforting about the myth of technological problem-solving when it is coupled with the fact of inequality. The term "structural" makes it seem as though "this is just the way it is," always a useful excuse for injustice.

Those who believe that intelligence tests mean something can verify the following theorem (based on the fact that test scores follow a Gaussian curve normalized so that the average is 100, with two-thirds of the population inside a standard deviation of plus or minus 15 points): whenever the intelligence requirement for a job goes up by one IQ point, then the number of people able to handle that job will go down by approximately one percentage point, and the unemployment rate will go up correspondingly, other things being equal. Even those who give little credence to IQ tests cannot fail to notice that as time goes on, more and more people become unfit (for whatever reason) to handle jobs that have become more and more demanding. The result is a stubborn residue of unemployables who present a serious problem to those of us who think that civilization should be able to deal with this problem in a better manner. Five to ten percent unemployed has been normal during the 1980s, whereas one to five percent was characteristic of the years earlier in the century (except for the thirties, the decade of the Great Depression).

If we look at reality squarely and dispense with the myth that all men and women are equal, we are free to expend greater efforts to educate and train those who are in danger of falling behind. It's cheaper in the long run than paying welfare and building jails. If it hurts the student's feelings to be moving more slowly than others, it will hurt his feelings even more to be unemployed and homeless.

Another myth that seems to be highly prevalent is the idea that "human nature is inborn." According to this myth, some people are good and some are evil, and there is not much you can do about it. If a person is evil and does bad things, put him in jail and throw the key away.

This is a topic so fraught with emotion and based on so little realistic knowledge that it would be foolhardy to pursue it in detail. The nature-nurture controversy has been argued to death and a conclusion is far from being reached. What does seem clear is that almost all high-level behavior is learned. I was born in the United States of Russian-Jewish parents. But I went to English-speaking public schools; as a result I speak English, not Russian. Isaac Asimov was born in Russia of Russian-Jewish parents, but he came to the United States at an early age; he speaks English also. Had he come to this country ten years later, he would be speaking English with a Russian accent instead of a Brooklyn accent. Clearly, where you were born does not determine which language you speak, nor does your heritage. There is no gene for language: it is purely a matter of programming, of education. But language is a very cerebral, high-level activity. What of lower-level activities?

We do not need education to learn how to breathe or to eat. That is built into the hardware. Whether you use a fork or a pair of chopsticks is determined by outside forces and comes with the software. We are told that spiders are born knowing how to spin a particular kind of web. This behavior is hardware-determined. The question we want to answer is: how complex can inherited behavior become? Or, using our computer metaphor, how much behavior can be translated into the hardware directly from the genes?

Observing the great variety and plasticity of behavior throughout the world, and observing how this behavior changes from year to year and from generation to generation, I am forced to believe that the idea of evil as inherent in humans is a myth. We observe that children abused by the adults around them grow up to be violent adults. It is programmed behavior. Does the process work in the opposite direction? Does less abuse produce better behavior? Sometimes, but not always.

It is probable that genetics determines general qualities of nervous response. The hardware specifies how fast the nervous system operates and how sensitive it is to both inner and outer stimuli. Observing children from the moment of birth, we see that some are immediately excitable, while others are placid from the start. It is clear that some children are going to grow up quickly prone to anger, aggression, and violence. However, the manner in which the anger is expressed will be determined by early programming and education.

It is also becoming clear that some elements of irrationality and mental illness are the results of bad wiring in the hardware. As Daniel Koshland, editor of *Science,* says, "Some inherent defects may be exacerbated by environmental conditions, but the irrational output of a faulty brain is like the faulty wiring of a computer, in which failure is caused not by the information fed into the computer, but by incorrect processing of that information after it enters the black box."[5] (Note how the computer model of human intelligence has already become a widely accepted metaphor.)

What are the realistic consequences of the myth that human nature is inborn and that evil is inherent in humanity? One consequence is that when the authorities see someone behaving badly, the immediate response is to put him in jail just to get him off the street. If the person is born bad, then there is no reason even to attempt rehabilitation in the prisons. According to this logic, there is no reason to improve bad neighborhoods or schools because the badness is built in and cannot be alleviated. Thus, children continue to grow in a bad environment, and as they grow up they continue to make the environment worse.

Belief in the myth that human nature is inborn results in a social system controlled by a classic feedback cycle: bad environment → bad children → bad adults → bad environment.

However, we see bits of evidence in our inner cities that harmful feedback cycles can be broken. It takes a concerted effort by the people living within the environment. It also requires help from the outside world—not only material help, but knowledge that shows what a better world is like. The myth of inherited evil is false. But belief in the myth and acting on that belief is evil.

Clearly, a new principle is in order: the *separation of myth and state.* Keep myths in the domain of pleasure. Enjoy them and enjoy the good feelings that they engender. But when it comes to action in the real world, reality must be the guide.

The United States has spent its wealth in defense against a mythical evil empire that turned out to be totally bankrupt, both economically and militarily. When the Soviet Union turned to reality and decided that it was best for them to give up the myth of communism and join the rest of the world, we suddenly discovered that their military might was in such disarray that there was no longer any point in maintaining an expensive defense against them. By that time we were bankrupt ourselves. Now we are so addicted to military expenditures that it will be hard to turn swords into plowshares. Myths can be damnably expensive.

NOTES

1. B. Bettelheim, *The Uses of Enchantment* (New York: Alfred A. Knopf, 1976).

2. J. Randi, *The Faith Healers* (Buffalo, N.Y.: Prometheus Books, 1987).

3. "Boston Jury Convicts 2 Christian Scientists in Death of a Son," *New York Times* (5 July 1990): p. A12.

4. Philadelphia Department of Public Health, *Vital Statistics Report,* 1982.

5. D. E. Koshland, Jr., "The Rational Approach to the Irrational," *Science* (12 October 1990): p. 189.

INDEX